The
Mimetic Brain

Studies in Violence, Mimesis, and Culture

Interdividual Psychology

W hat is remarkable in my view is to have witnessed in the course of a lifetime the birth of a psychological and anthropological theory and to see it resonate about thirty years later with neuroscientific research. When René Girard published his theory of mimetic desire in 1961, Andrew Meltzoff's experimental research had yet to begin and mirror neurons wouldn't be discovered until the 1990s.

In 1978, in *Things Hidden since the Foundation of the World*, René Girard, Guy Lefort, and I forged the expression "interdividual psychology" to express our conviction that the monadic subject doesn't exist, that the self is formed only in relations with the other, and that psychology cannot focus on individuals but only on rapports and relationships: an interdividual psychology. All of these elements converging from several different anthropological and scientific sources led us—and in some sense obliged us—to rethink anthropology and psychology in a radically new manner. This is what I proposed to do in 1982 in the French version of *The Puppet of Desire*, and the theories that I advanced elicited the enthusiasm of only a few psychologists and philosophers, mostly in the United States, without attracting wider attention. I propose to revisit these texts and to reread them in light of the recent discoveries in experimental psychology and neuroscience.

From Mesmer's Fluid to "Mimesis"

It seemed to me that mimetic desire—which remained invisible in the normal gestures of life and in "normal" situations from day to day—could be coaxed into the open by the comparative study of psychological and psychopathological phenomena such as hypnosis, African possession or adorcism, diabolical possession and exorcism, and finally hysteria. These phenomena, illuminated by mimetic perception, seemed to me to complete and clarify one another.

Starting in 1981, I was struck by the question of why the child imitated the adult. Meltzoff had just taught us that imitation was the first mode of entering into relation with the other and that this imitation, through its later development, would enable the child's evolution and learning. I then postulated the existence of a heretofore unidentified force that would in a certain sense oblige the child to imitate the adult's gesture or to repeat the phonemes that the adult pronounced, leading to language acquisition. I suggested that this force be called *mimesis*. This mimesis was, to be sure, imitation in space, that is to say the imitation of apparent gestures, but it seemed to me that it extended across time through repetition, which enabled not only language learning but little by little deferred representation and thus the gradual constitution of memory. Finally, I was bold enough to imagine that the same mechanism was at work in the species to ensure reproduction.

From this starting point, I formulated the hypothesis that there existed in the human race a force that I called "universal mimesis" that was imitation in space, repetition in time, and reproduction in the species. I made the connection with the theories elaborated by Franz Anton Mesmer in the eighteenth century. This Viennese doctor, who practiced hypnosis, had observed in a clinical setting that there was at work between humans a force of attraction or repulsion that played a fundamental role in the contagion and propagation of ideas and feelings. Mesmer's problem was that he wanted to assert the existence of a universal "fluid" deriving from the stars and flowing through all human beings, linking them to one another; he went still further by explaining pathology by the magnetic constitution of this fluid, which he called "animal magnetism" and which, like mineral magnetism, could explain both the attraction and repulsion of two poles according to the way they were presented to each other. King Louis XVI's

The
Mimetic Brain

Jean-Michel Oughourlian

Translated by Trevor Cribben Merrill

Michigan State University Press · *East Lansing*

Copyright © 2016 Michigan State University Press; Notre Troisieme Cerveau:
La nouvelle révolution psychologique © Editions Albin Michel—Paris 2013

⊖ The paper used in this publication meets the minimum requirements of ANSI/NISO
Z39.48-1992 (R 1997) (Permanence of Paper).

 Michigan State University Press
East Lansing, Michigan 48823-5245

Printed and bound in the United States of America.

22 21 20 19 18 17 16   1 2 3 4 5 6 7 8 9 10

LIBRARY OF CONGRESS CATALOGING-IN-PUBLICATION DATA
Oughourlian, Jean-Michel.
[Notre troisihme cerveau. English]
The mimetic brain / Jean-Michel Oughourlian ; translated by Trevor Cribben Merrill.
pages cm. — (Studies in violence, mimesis, and culture)
Includes bibliographical references and index.
ISBN 978-1-61186-189-1 (pbk. : alk. paper) — ISBN 978-1-60917-476-7 (pdf) — ISBN
978-1-62895-247-6 (epub) — ISBN 978-1-62896-247-5 (kindle) 1. Neuropsychology.
2. Cognitive neuroscience. I. Title.
QP360.O8413 2016612.8'233—dc23
2015022381

Book design by Charlie Sharp, Sharp Des!gns, Lansing, MI
Cover design by David Drummond, Salamander Design, www.salamanderhill.com.
Cover art: Reversible Head with Basket of Fruit (c. 1590) by Giuseppe Arcimboldo
(Arcimboldi) (c. 1526–1593).

🌱 green press INITIATIVE Michigan State University Press is a member of the Green Press
Initiative and is committed to developing and encouraging ecologically
responsible publishing practices. For more information about the Green Press Initiative
and the use of recycled paper in book publishing, please visit www.greenpressinitiative.org.

Visit Michigan State University Press at www.msupress.org

Contents

vii   Author's Note

ix   Preface

xvii   Introduction

PART 1. TOWARDS A MIMETIC ANTHROPOLOGY

3   CHAPTER 1. Contagious Desire

17   CHAPTER 2. The Precursors

23   CHAPTER 3. Some Contemporaries

PART 2. A PSYCHIATRY OF THE THREE BRAINS

33   CHAPTER 4. Interdividual Psychology

39   CHAPTER 5. Psychological Time and the Nodal Points N and N'

49   CHAPTER 6. The Three Brains

57   CHAPTER 7. The Three Possibilities of the Interdividual Rapport

65   CHAPTER 8. Classical Nosology

PART 3. AN ESSAY IN MIMETIC NOSOLOGY

73   CHAPTER 9. Figures of the Other in Normal Experience

89   CHAPTER 10. Figures of the Other in Neurotic Experience

103 CHAPTER 11. Figures of the Other in Psychotic Experience

133 CHAPTER 12. Mood Disorders

143 CHAPTER 13. Diseases of Desire

PART 4. THE DIALECTIC OF THE RATIONAL, EMOTIONAL,
AND MIMETIC BRAINS

165 CHAPTER 14. The Mimetic Mechanism

171 CHAPTER 15. Some Clinical Studies

187 Conclusion

201 Notes

211 Bibliography

215 Index

Author's Note

The mechanisms brought to light in this book will appear more and more clearly in proportion as our Western and Judeo-Christian civilization is destructured and, in a certain sense, collapses. In the eighteenth century and even in the nineteenth, the separation between "normal" and mentally "healthy" people on the one hand and the "alienated" and the mad on the other ("alienated" meaning "having become foreign to the community") was clear and admitted no debate. And normal people had, in principle, no psychological problems. It was with Charcot, Bernheim, and above all Freud that we began to evoke the psychological problems of people who were relatively normal—or in any case not locked up—but these problems remained in a sense inherent and interior to the individual, the subject. It seems to me that as our culture disintegrates over time, the "interdividual" mechanisms, and the rivalries they generate, will appear universal. The distinction between normal people on the one hand and the "mad" on the other will tend to disappear. The problems are generalized and spreading, and this book will attempt to show to what degree psychopathology is becoming interdividual and relational.

Preface

From the beginning of my medical studies in Strasbourg, and then of my studies in psychiatry at the Sainte-Anne Hospital in Paris, I became aware that the ideas my teachers and elders had about human beings, about the structure of their psychic apparatus, and about the greater or lesser role that could be attributed to their surroundings, environment, personal history, and culture, greatly influenced their practice, their way of treating patients, and their psychotherapeutic approach.

Fifty years have gone by and I have encountered various and varied theories and practices. I have personally been led to change my anthropology, and my ideas about the mental and psychic architecture of human beings and about the determining factors in their behavior have evolved a great deal, gradually enriched by my encounters, the books and articles I have read, my research, my psychiatric medical practice, and the revolutionary scientific discoveries that have marked the last fifty years in my field.

The first half of the twentieth century was dominated, in psychology as in psychopathology, by the problem of consciousness.[1] Pierre Janet defended the idea that psychopathological problems, mental illnesses, neuroses, and psychoses were due to a "shrinking of the field of consciousness." When one is obsessed by an idea, a fear, or a person, the field of consciousness shrinks and concentrates on this object, losing sight of all others and thus of the freedom of thought. Nonetheless, Janet also affirmed the existence of the subconscious, the clinical reality of which he demonstrated thanks to numerous experiments in hypnosis carried out at Salpêtrière Hospital in Paris, notably concerning the phenomenon of posthypnotic suggestions.

Freud's theory and his "discovery" of the unconscious swept away Janet's theories, and when I arrived at Sainte-Anne, psychoanalysis dominated the human sciences within academia. Freud's "topics"[2] were accepted as translating the reality of the psychic apparatus's structure. Each psychoanalyst, however, had his own reading of Freud, and these various approaches were very different from one another. The most eccentric was that of Jacques Lacan, whose seminar I attended a few times without understanding a thing. And in any case, in a general way, psychoanalytic theory appeared "mythical" to me. Schooled in Greek and Latin culture, I knew that the ancients explained everything that happened to human beings by the interplay of conflicts and rivalries among the various gods who resided on Olympus and whose benevolence or anger explained everyone's destiny. It seemed to me that, in the same way, the unconscious was like a sort of inaccessible Olympus, where mythical entities, namely the id, the superego, Eros, Thanatos, the libido, and the various instincts and drives, engaged without our knowing it in rivalries and combats the result of which was our behavior, our destiny, our joys and our sufferings. Faced with this anthropology, I was perplexed, all the more so because I saw its beneficial effects on certain therapists who, by relying on this theoretical framework, had found a balance and self-assurance that helped their patients greatly.

Generally speaking, then, the conscious-unconscious dialectic dominated the human sciences and in particular psychology and psychopathology.

The problem of consciousness was equally central for my teacher Henri Ey, who at that time had been baptized "the pope of French psychiatry" and who twice weekly, at Henri-Rousselle Hospital and at the Sainte-Anne library, gave seminars and presentations of sick patients. For Henri Ey, mental illnesses were the result of a "destructuration of consciousness," a destructuration that was more profound the more serious the illness was. He called his anthropological theory "organo-dynamism."

Henri Ey taught me semiology (the classification of symptoms) and nosology (the classification of syndromes and illnesses) and how to make diagnoses in psychiatry. He taught me how to approach patients, whether neurotic or psychotic. By his example and his teaching, he exorcised my fear of mental illness and replaced it with scientific and human interest in the patient.

Henri Ey was a man of the South, a Catalan who was so solid, so jovial, so knowledgeable, and so sure of himself that his personality won you over. His benevolent attitude toward me made it possible, during private conversations, for me to observe that the destructuration of consciousness didn't exist in a psychosis as well documented as paranoia and that it was, to the contrary, very significant in the "queen of neuroses," hysteria. He laughed and encouraged me to seek out an answer.

On the other side of the Sainte-Anne Hospital, I was an intern in the ward of my teachers Jean Delay and Pierre Deniker. A few years before, with the discovery of the antipsychotic effects of Largactil, they had inaugurated the era of psychopharmacology, which was to experience a spectacular development in the second half of the twentieth century. It was they who established the classification of psychotropic drugs, with psycholeptics (which diminish psychic activity: tranquilizers and neuroleptics), psychoanaleptics (which increase psychic activity: amphetamines and antidepressants), and finally psychodysleptics (which deviate psychic activity: hallucinogenics, essentially). It was also they who, supported in their city practice by my teacher and friend Louis Bertagna, introduced lithium as a treatment for "mood disorders."

My experience in Delay and Deniker's ward, under the supervision of my teacher and friend, Dr. Georges Verdeaux, taught me how to administer psychotropic medications. The latter gave me the feeling that I was a bit of a "sorcerer's apprentice," for when faced with a symptom (depression, delusion, hallucinations) I knew what medication to prescribe and in what dose, yet without for all that knowing how it worked! It was an empirical and, if only in the sense that it was based on experience, a scientific practice. Today we know much more about the biochemistry of neurotransmitters, but most of the time we are still sorcerer's apprentices.

On a theoretical level, the daily care of patients distanced me from the conscious-unconscious issue. That is when I became aware of a problem that had been around for several centuries: Cartesian dualism. How should we think about a soul-body dualism if a chemical substance (Valium, for example) introduced into the body could have such a direct effect on the "movements of the soul," on anxieties, fears, metaphysical reflections, which disappear as if by enchantment once the injection has taken effect? How could the corporal, the physical, have an influence on the psychic, on the soul, when soul and body were not of the same nature?

I came upon the problem from the opposite angle when, several years later, under the supervision of my teacher and friend Aimé Burger, I discovered psychosomatic medicine. The latter was very much in fashion at the time and gave rise to an abundant literature. But, once again, how could Cartesian thought integrate the influence of psychic torments, of feelings, in short of movements of the soul on the body, the soma, producing functional pathologies like aches and pains, headaches, nausea, high blood pressure, and so on, or even pathologies caused by a lesion like stomach ulcers, Crohn's disease, and so on? All of this posed a theoretical problem that also merited reflection and research, and that Jean Delay summed up admirably: "Psychic disorders organize themselves and little by little organicize themselves."

Psychological and psychopathological problems were thus also anthropological and philosophical problems. So I headed for the Sorbonne and the human sciences, where I encountered eminent men who broadened my horizons in an astonishing way: teachers who became friends like Henri Faure and Claude Tresmontant; distant but exciting teachers like Paul Ricoeur, Emmanuel Levinas, and Mircea Eliade. I was also initiated into the thought of Heidegger, Saint Thomas Aquinas, Aristotle, and Plato, and I deepened my understanding of Freud's theory of interpretation.

But the encounter that was to orient the rest of my life and my research was the one with René Girard in 1972. I was finishing my doctorate on drug addiction and I was looking in vain for a book that could enlighten me about the problem of violence, which was at the heart of my clinical research at the time. My bookseller, Monsieur Gizard, brought me a book one day while I was sick in bed with a high fever. I emerged from my stupor a few days later and found the book on my bedside table. It was *Violence and the Sacred*. I devoted the months of October, November, and December 1972 to reading that incredible work four times over. I decided to meet René Girard as soon as possible to express my admiration and to tell him that I thought his theories could apply to psychology and psychiatry. It was in spring 1973 that I was able to go to Buffalo, New York, where René was teaching at that time.

I understood immediately that René Girard was a pure genius and that one of his numerous intellectual qualities was his sense of humor. Later, when we were working together on *Things Hidden since the Foundation of the World*, our lack of reverence for the great thinkers was absolute. Shared laughter gave us both energy and encouraged René, I think, to write boldly things that perhaps he would otherwise have hesitated to write. I would like to think that I helped René because I was an enthusiastic interlocutor, but also because ideas become clearer and more precise when they are explained to an interested student. I also think that I convinced him that his theory could revolutionize psychology and psychiatry, and it was for this reason that we decided to draft the third part, to which we gave the name of the new psychology that we wanted to promote: "Interdividual psychology." This terminology sought to evacuate the notion of the individual, of the monad, the subject, and to replace it with a psychology grounded essentially in the mimetic relationship and the interdividual rapport, that is to say, more generally, in the *mimetic reciprocity* whose principal developments I will be addressing in this book.

I consider it a privilege to have had René Girard as a model who, by his example, and through the work we did together, and thus in a certain sense without his realizing it, taught me several things. He taught me to read, that is to say to dive into another author's thought with both interest and a distance that made it possible to conserve one's freedom, and enabled one to use a great thinker's text as a springboard for personal reflection. He gave me what I would call "mimetic spectacles" enabling me to see all around me, in daily events as well as in the texts of world literature, hitherto invisible realities.

Above all, he gave me the most invaluable gift: his friendship, his time, and the opportunity to work with him. René subsequently gave me two other gifts. The first was introducing me to his dear friend Michel Serres, who honored me in turn with his friendship, a friendship that for forty years has been very precious to me. Michel offered me his advice while I was preparing my *thèse d'État* and was the head of my thesis committee. And it was he who, as we were driving through the streets of San Francisco with René Girard, told me: "Look at that streetcar in front of us. It reminds me of *A Streetcar Named Desire*. Why don't you call your next book *Un mime nommé désir* [3] (*A Mime Named Desire*)?" The second gift that René gave me was introducing me to Benoît Chantre, president of the Association Recherches Mimétiques and a fellow of the Imitatio Foundation in the United States. Benoît's stimulating friendship helped me to come back to writing, and he "coached" me as I wrote and published *The Genesis of Desire* in 2007.

◆   ◆   ◆

Two fundamental hypotheses structure René Girard's entire body of work, namely the psychological hypothesis of mimetic desire and the sociological hypothesis of the scapegoat mechanism, which enabled humans to survive their own violence via the organization of a system made up of myths, rituals, and taboos or prohibitions. Those three pillars, on which the religious mechanism is built, issue directly from the immolation of the primordial emissary victim, at the foundation of culture. The enormous advantage that I found in this way of thinking—its purity and capacity for generating complexity filled me with enthusiasm—was that, like Ulysses's arrow passing through all the rings set up by Penelope, René's theory proved capable of illuminating each and every discipline in the human sciences in turn.

From the time of my encounter with René and my mimetic initiation, my whole philosophical and anthropological perspective was modified, and I began reading psychology and psychiatry in light of these new ideas. Moreover, I was interested henceforth in everything connected in some way to mimetic studies, in psychology as well as in neuroscience. This book is an attempt to take stock of all those stages and all those encounters, and of the results obtained from fifty years of clinical research and practice.

◆   ◆   ◆

Each time that I receive a new patient, I have the same feeling as a card player at the beginning of a game. The cards are always the same. There are four aces of various colors, and so forth, and no more. But the hand we are dealt, the game we are about to play—these are always different. And for years and years, all over the world, with the same fifty-two cards, billions of games of bridge and poker have been played without any of them being exactly like the others. The same goes for the games of mimetic desire. In this book I will seek to spot the cards, which are always the same, instead of remaining prisoner of each game's singularity. In this sense, it's an approach that can be thought of as scientific, and which, in any case I hope, will make it possible—especially for young psychiatrists—to have better insight into psychopathological situations.

I am aware that each human being is unique, and that each case is particular, that generalizations and systematizations are necessarily schematic and that, as a result, they necessarily amputate a part of reality. Nevertheless, it is impossible for a young psychiatrist to find the courage to treat a mentally ill person, a neurotic or a psychotic, unless he has a template that makes it possible to make sense of what he is seeing, and thus to feel reassured and able to be effective. For if he lacks this anthropological underpinning for his therapeutic activity, the behaviors he observes will appear to him absurd, threatening, dangerous, and incomprehensible, and thus as a source of anxiety. And the anxiety he will feel when faced with this strangeness will paralyze his therapeutic activity.

Indeed, what is important for a young psychiatrist is having references that make the phenomena he is observing intelligible. This is reassuring, it diminishes his anxiety, and it makes him perform better. Throughout history, those who sought to create a template for interpreting psychological phenomena have contributed to training psychiatrists. Freud, Janet, Charcot, Bernheim, Mesmer, Jung, Lacan, Henri Ey, and so on—they all put forth good ideas. When, during my years of training, I discovered the thought of one of these masters and spoke of it with Georges Verdeaux, he would say to me good-naturedly: "Yes, you can see things that way." Today, he would probably say the same of this book. In other words (to return to my playing card metaphor), not only are the hands and the games different every time, but the gaze we turn on the scenes before our eyes is always true in a certain sense and incomplete in another. That is why this book offers a new way of seeing things, which will no doubt be as incomplete as the others, but will also contain a portion of the truth in sufficient quantity, I hope, to give young psychiatrists an alternative theoretical foundation for their clinical activity.

Introduction

O bviously, each human being possesses only one brain. And in reality, it is each brain that "possesses" and constitutes a human being. And yet, on an anatomical and historical level, various structures have been recognized as responsible for particular functions, contributing to the overall working of the brain, and as a result, of the individual.

In the past, scientists and philosophers thought that judgment and the capacity to choose were determined above all by formal reasoning, which was considered to be the primary task of the thinking cortex. The emotions were phenomena that had to be tempered and which, as far as rational behavior was concerned, were treated with suspicion. Ontology and thought were very closely related. This is the basis of the Cartesian approach: "I think, therefore I am." The cortex was also the seat of memory. Freud would later assert that the memory is divided into a conscious part and a distinctly separate and unconscious part, whose role is a bit like the Greek Olympus where the destinies of men were decided. Throughout this book, in a schematic way, I will call this cortical, rational brain the *first brain*, a term justified on both the historical and the anatomical levels, since neurologists discovered it first, in particular by isolating the zones of motor function, sensitivity, language, and the five senses.

A first revolution in neurology and psychiatry took place with the discovery by Antonio Damasio of the fundamental importance of another part of the brain, which had been neglected until then, the archaic brain, which some still called the "reptilian brain," represented by the limbic system, which Damasio baptized the "emotional brain." This was a very important discovery that corresponded to highlighting the fundamental activity of all the cerebral zones gathered under the name "limbic system," which I propose to call the *second brain* here.

Following Damasio's publications, we were flooded with publications on emotional intelligence and it was shown that the cortex (what I am calling the "first brain"), without a permanent symbiosis with the second brain, could not account for an individual's normal behavior. The limbic system had to be called upon.

At this stage in the evolution of thinking about the brain, the first and second brains, that is to say the cortex and the limbic system, seemed to be the two sole drivers of all psychological movement. Rationality and emotionality formed a couple whose agreements and conflicts, harmonies and disharmonies, appeared to explain in and of themselves all the figures of normal and pathological psychology.

But it seems to me that a third entity plays a decisive role in our psychological makeup. It is time to expand our vision of the latter by introducing this new variable, which is rarely mentioned but whose role is capital, namely *relationship, reciprocity, mimeticism.* We have known, indeed, since the mid-1990s that human beings are equipped with a neuronal apparatus that negotiates relations with the other on the most basic level, before thoughts and emotions are called upon: the mirror system.[1] The mirror system puts humans in a prerational resonance with other humans. It makes the "individual" capable of identifying the other's gestures, of interpreting the other's actions and intentions and of understanding and imitating them, and may even constitute the underlying neurophysiological foundations of empathy as well as explaining the determining factors in human action.

So-called mirror neurons, which were discovered first in monkeys but which have been identified and localized in the human brain as well, are anatomically a part of both the rational and the emotional brains, but the relations that they make it possible to establish with the brains of other human beings have such importance and such a psychological reality that this mimetic interdividuality (I will explain this term in detail later on) deserves in my view to be labeled as the *mimetic brain* or *third brain*.

One of the central arguments that I put forward in this book is that this relational function—which is essentially imitative—is the driving force of the emotional and cognitive functions. Before making use of our rational capacities, we absorb much more immediately and mimetically information to which our mirror system gives us access. With the hypothesis of a "shared manifold," the neuroscientist Vittorio Gallese has emphasized this "shared network" of mirror neurons. He invites us to think about the interrelations and imbrications of the mirror neuron system, the affective system, and the most conscious level of personal experience, that is to say thought and reflection: "Sensations, pains and emotions displayed by others can be empathized, and therefore understood, through a mirror matching mechanism," he writes in a seminal 2001 paper.[2] He also hypothesizes that higher-order brain functions may depend on these basic mirroring mechanisms: "The discovery of mirror neurons in the monkey premotor cortex has unveiled a neural matching mechanism that, in the light of more recent findings, appears to be present also in a variety of non motor-related human brain structures."[3] In his conclusion, Gallese states that human thinking and reasoning abilities may even depend upon these prerational, prelinguistic simulations: "My conclusion is that the more we'll know about how our brain-body system works, the less remote the nature of thought and reasoning will appear from it."[4]

Gallese grounds our emotion and cognition in basic interrelational mechanisms. I think that this new way of seeing things makes possible a revolution in psychopathology, based on the notion of imitation and, most importantly, of *mimetic rivalry*. Obviously, however, I am not a neuroscientist. I am a clinician inspired by the mimetic theory of René Girard, and my metapsychology was first published in *The Puppet of Desire* in 1982. So the reader must realize that the discovery of mirror neurons appeared to me as a new clue pointing to the same conclusions and not at all as the beginning of my new outlook on psychology and psychopathology.

The new paradigm that I am trying to develop integrates recent discoveries in neuroscience and developmental psychology with the existing models. We have a cortical brain that brings us rationalizations and justifications that can be economic, political, moral, religious, and so on. To this system is added a limbic system that governs our emotions (stress, anxiety, anger, joy, fear, and so forth), our feelings (love, tenderness, hatred, envy, jealousy, resentment, and so forth), and our moods (euphoria, excitation, depression, lethargy, and so forth). Until now, psychiatry and psychology have concentrated either on the analysis of the contents of the cortex—that is to say thoughts, morality, religious or economic justifications—or on moods, feelings, and emotions. Until very recently, nobody was aware of the mirror system's existence, and nobody, or almost nobody, was interested in imitation. And when René Girard, who was the first thinker of the twentieth century to look into these questions and to connect imitation and rivalry (this Gabriel Tarde did not do), spoke about them in his books, virtually nobody wanted to see that before anything else there was the imitation of others, and that this imitation would determine the tenor of our feelings and thoughts.

◆   ◆   ◆

In my approach to psychopathology, the reader will find many similarities and probably many differences with respect to the various approaches of relational psychoanalysts, who emphasize to varying degrees the importance of the relationship between people and between the analyst and the analysand. It would be impossible to list all the psychoanalysts since Freud who have engaged in such an undertaking. Let us take as representative the very interesting synthesis attempted on the subject by Stephen A. Mitchell in his book *Relational Concepts in Psychoanalysis*.[5] Particularly striking are Mitchell's comments on Bowlby's attachment theory, which he outlines as follows: "The first general strategy for addressing the question of the origin and motivations of personal relatedness might be characterized by the answer: *because we are built that way*. People are constructed in such a fashion that they are inevitably and powerfully drawn together."[6] This is just one example among many showing that all psychologists and psychiatrists since Mesmer in the eighteenth century have been fascinated by the obvious attraction of human beings to one another (indeed, in some respects, Mesmer's insights were superior to those of attachment theory, for despite their pseudoscientific air they sought to explain not only attraction but also the repulsion that may arise between two people).

One of the main differences between the various theories of relational psychoanalysis and the theory presented here is that relational psychoanalysis assumes that the relationship is built, engineered, and shaped by two subjects

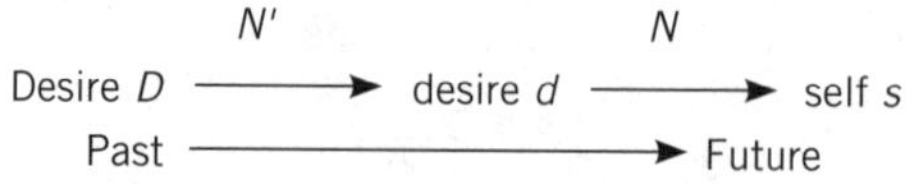

$$\text{Desire } D \xrightarrow{\quad N' \quad} \text{desire } d \xrightarrow{\quad N \quad} \text{self } s$$
$$\text{Past} \xrightarrow{\hspace{4cm}} \text{Future}$$

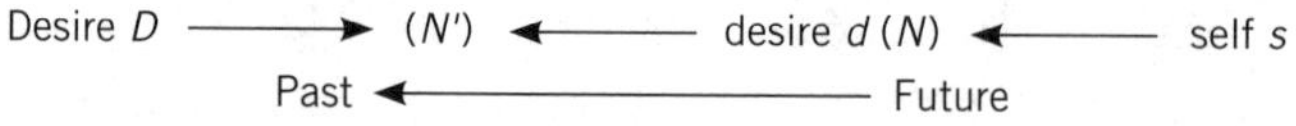

$$\text{Desire } D \xrightarrow{\hspace{1.5cm}} (N') \xleftarrow{\hspace{1.5cm}} \text{desire } d\,(N) \xleftarrow{\hspace{1.5cm}} \text{self } s$$
$$\text{Past} \xleftarrow{\hspace{4cm}} \text{Future}$$

Spinoza and that finally offers a very important confirmation of René Girard's theory: learning is the result of mimeticism and of mimetic desire when the model remains a model; the pupil wishes to imitate the words that the model repeats to him or her, and the latter wishes the pupil to learn and to appropriate them. It is the same mechanism that, bearing on an object that the model designates but withholds, will lead to rivalry and conflict.

· Finally, we must emphasize the fact that the observer's mirror system reflects the intention of the action he is witnessing, even if it is not completed. "Hooked up" to the same wavelength, so to speak, the observer's brain guesses the other's intention, that is to say desire, and models him- or herself on it, even if the gesture is unfinished or if the hand reaches for a hidden object (a piece of food placed behind a screen and that the subject thus does not see when the experimenter's hand plunges behind the screen).

It remains to be seen whether mirror neurons, in the cognitive, motor, sensory, and sensitive cortex, as well as in the limbic brain, are specific and individualized anatomical structures, or whether almost all neurons have a "mirror function" that is activated in the relationship with the other. Research on this question is under way, but I won't hide my personal intuition that the second hypothesis will one day be verified. It is also possible that both are founded: there may be purely mirror neurons and others having, in addition to their specific function, a "mirror function."

A Psychiatry of the Three Brains

CHAPTER 4

Interdividual Psychology

W hat is remarkable in my view is to have witnessed in the course of a lifetime the birth of a psychological and anthropological theory and to see it resonate about thirty years later with neuroscientific research. When René Girard published his theory of mimetic desire in 1961, Andrew Meltzoff's experimental research had yet to begin and mirror neurons wouldn't be discovered until the 1990s.

In 1978, in *Things Hidden since the Foundation of the World*, René Girard, Guy Lefort, and I forged the expression "interdividual psychology" to express our conviction that the monadic subject doesn't exist, that the self is formed only in relations with the other, and that psychology cannot focus on individuals but only on rapports and relationships: an interdividual psychology. All of these elements converging from several different anthropological and scientific sources led us—and in some sense obliged us—to rethink anthropology and psychology in a radically new manner. This is what I proposed to do in 1982 in the French version of *The Puppet of Desire*, and the theories that I advanced elicited the enthusiasm of only a few psychologists and philosophers, mostly in the United States, without attracting wider attention. I propose to revisit these texts and to reread them in light of the recent discoveries in experimental psychology and neuroscience.

From Mesmer's Fluid to "Mimesis"

It seemed to me that mimetic desire—which remained invisible in the normal gestures of life and in "normal" situations from day to day—could be coaxed into the open by the comparative study of psychological and psychopathological phenomena such as hypnosis, African possession or adorcism, diabolical possession and exorcism, and finally hysteria. These phenomena, illuminated by mimetic perception, seemed to me to complete and clarify one another.

Starting in 1981, I was struck by the question of why the child imitated the adult. Meltzoff had just taught us that imitation was the first mode of entering into relation with the other and that this imitation, through its later development, would enable the child's evolution and learning. I then postulated the existence of a heretofore unidentified force that would in a certain sense oblige the child to imitate the adult's gesture or to repeat the phonemes that the adult pronounced, leading to language acquisition. I suggested that this force be called *mimesis*. This mimesis was, to be sure, imitation in space, that is to say the imitation of apparent gestures, but it seemed to me that it extended across time through repetition, which enabled not only language learning but little by little deferred representation and thus the gradual constitution of memory. Finally, I was bold enough to imagine that the same mechanism was at work in the species to ensure reproduction.

From this starting point, I formulated the hypothesis that there existed in the human race a force that I called "universal mimesis" that was imitation in space, repetition in time, and reproduction in the species. I made the connection with the theories elaborated by Franz Anton Mesmer in the eighteenth century. This Viennese doctor, who practiced hypnosis, had observed in a clinical setting that there was at work between humans a force of attraction or repulsion that played a fundamental role in the contagion and propagation of ideas and feelings. Mesmer's problem was that he wanted to assert the existence of a universal "fluid" deriving from the stars and flowing through all human beings, linking them to one another; he went still further by explaining pathology by the magnetic constitution of this fluid, which he called "animal magnetism" and which, like mineral magnetism, could explain both the attraction and repulsion of two poles according to the way they were presented to each other. King Louis XVI's commission formed to look into the matter (Franklin, Lavoisier, Bailly, and others), after having studied the magnetic techniques and theories of Mesmer, concluded that there existed no such universal fluid and that pure magnetism had no effect without suggestion, but that suggestion without magnetism was demonstrably effective.

Echoing Mesmer, Leonhard Euler, in a letter to a German princess in 1772, wrote: "But when we want to penetrate the mysteries of nature, it is very important to know if the celestial bodies act upon one another by impulsion or attraction; if some subtle and invisible matter pushes them against one another, or if they are endowed with a hidden or occult quality by which they are mutually attracted."[1]

Where the similarity between the phenomena of attraction and repulsion of celestial bodies and human beings was perceived by Mesmer and attributed to a magnetic fluid, Euler spoke of a "subtle matter" or of a "hidden or occult quality." In fact, by formulating a physical hypothesis, all of these authors were seeking to explain the psychological reality of the mimetic interdividual rapport represented by the back-and-forth of imitation and suggestion between human beings.

Toward a New Metapsychology

This led me to develop two ideas.

The first sprang from the comparison with the planets and the astrological system, which I found interesting. I told myself that the universal gravitation discovered by Newton governed the physical realm and that universal mimesis governed the human realm according to mechanisms that could be comparable, making a unified metaphysical conception of the universe possible. This comparison is obviously speculative and metaphorical, but it seemed to me enlightening as a means of understanding the way psychosociology works. Universal gravitation of course explains the attraction that celestial bodies exert on one another. The question was then: how do these celestial bodies avoid crashing into one another? The answer was by means of movement. If the moon doesn't crash into the earth, this is because the movement that it makes around the earth keeps it separated while at the same time obliging it not to move farther away. It seemed to me that in psychology universal mimesis could lead humans to "crash into" one another and to fuse into a sort of coalescence, but that what prevented them from doing this was movement. In psychology this movement that obliged them to orbit around one another, to alternatively come closer or pull back, seemed to be desire. Indeed, the adult's hand, moving toward an external object, pulls the child's desire and movement toward that object, even as the child follows a centripetal trajectory that distances it from the adult's body. And in this way the child is gradually led to detach itself from the maternal breast and from a conjoined relationship as objects of desire that pull it away from the mother's body are mimetically suggested to it. Thus, in psychology, desire is movement and movement is desire.

The second idea was based on the fact that Louis XVI's commissioners had ascribed to suggestion hypnotic phenomena and the contagion of ideas and feelings that Mesmer explained by means of animal magnetism. It seemed obvious, then, that between two psychological entities, which I called "holons"[2] to avoid calling them "subjects," there existed two vectors, one from A to B that was a suggestion but that could only exist if B was imitating A, that is to say if there was a vector going in the opposite direction but following the same trajectory, an imitative vector.

Human interaction is based on this principle of reciprocal imitation. Two people meet. One of them holds out his hand. The other imitates him and holds out his hand in turn. Now they are on friendly terms, thanks to positive imitation, good reciprocity. (We all keep in our memories the famous handshake between Arafat and Rabin, watched over by the president of the United States to publicize and emphasize the new friendship that he had brought about between Palestinians and Israelis.) But if the second rejects the proffered hand, the first gets angry and says (for example): "Go to hell." Now they are enemies, caught up in rivalrous, "bad" reciprocity. The second refused to imitate the first one's gesture, and the first one immediately imitated the second one's hostile attitude. The reader can of course refer to René Girard's book *The One by Whom Scandal Comes* for further discussion of reciprocity.[3]

In reality, the first one's extended hand is a suggestion, which is supposed to entail the second one's imitation. This relational reciprocity is mimetic in its essence, and it is universal. The interdividual rapport can be represented in figure 5. The two vectors are identical, they respond to each other, and, fixed in this way, they represent good reciprocity. The two vectors are multiple and they come and go "cinematographically" between A and B.

Figure 5.

In the case of bad reciprocity, A's suggestion will be rejected by B and A will hasten to imitate this rejection, in other words to imitate in turn B's negative suggestion (see figure 6). I then concluded from this that imitation and suggestion constituted the back-and-forth of the rapport between two humans that I called, after working with Girard on *Things Hidden*, the "interdividual rapport." This interdividual rapport expressed the fact that there were not two individuals isolated from each other but rather a perpetual movement and imitation-suggestion vectors circulating in a cinematographic and not a photographic manner as in the "frozen" image above.

Figure 6.

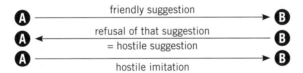

At once, hypnosis—to take one example—became clearer: in this phenomenon, the interdividual rapport was—thanks to the force of the suggestion vector going from the hypnotizer to the hypnotized—immobilized in a certain sense, thus fixing the vector going from the hypnotized subject to the hypnotizer in the imitation mode. At this instant, the interdividual rapport between the hypnotizer and the hypnotized was stabilized, abandoning the usual back-and-forth proper to a normal interdividual rapport and making the imitation of the hypnotizer's desire irresistible for the hypnotized subject through the immobilization of the two vectors—suggestion in one direction and imitation in the other. From there, drowsiness, followed by lethargy and submission of the hypnotized subject's self could be observed. But under the influence of the hypnotizer's continuous suggestion, a new self was seen to appear, one that was obviously formed and modeled by the hypnotizer's desire, this new self appearing with all its newly formed attributes: a new consciousness, a new memory, a new sensibility, a new emotivity.

And from there I grasped the necessity of a new metapsychology in which desire would no longer be the self's desire, as had always been thought, but where the self would be the self-of-desire. And as desire is mimetic, the desire forming this self would be in fact the reflection, the copy of the other's desire. Later on, Eugene Webb would suggest giving to this self generated by interdividuality an evocative name, the "self between."[4]

In a clinical setting and in all of my conversations with patients, I observed that a double claim kept being made by all of them, not only neurotic or psychotic patients but also normal ones: the claim by the self to the ownership of "its" desire and the claim by desire of its anteriority and priority with respect to the other's desire, which had in fact engendered it through mimetic suggestion. This suggestion could be either deliberate or inadvertent on the model's part. I then imagined that this new metapsychology could bring psychology into a scientific space, since we had two constants that could be more or less easily identified in humans: a nodal point N representing the self's claim to the ownership of its desire and a nodal point N' representing desire's claim to anteriority over the other's desire, which was responsible for inspiring and generating it.

It remained to study how normal psychology was reflected and manifested in N and N', and what were the neurotic and psychotic strategies used respectively by the self and by desire in these same nodal points to lay claim to their double pretention. It seemed to me that this was the object of a new psychological and psychopathological approach to clinical realities.

Psychological Time and the Nodal Points N and N'

How can we imagine what happens at nodal points N and N'? The self, at point N, in the most banal and normal case, cannot survive unless it is persuaded that it is the owner of its desire. The simplest solution for the self consists in forgetting the otherness of the desire that constituted it and in considering that this desire truly belongs to it. In reality, it is not a matter of mere forgetting because if one forgets something, this implies that one once knew it. It is in fact a matter of active misrecognition, though at this stage remaining peaceful and nonadversarial.

The constitution of the self in physical time can be summed up by a linear vector going from the past toward the future. Desire D mimetically elicits the birth of desire d, which, in turn, brings self s into existence. Such is the real sequence of events that unfold in physical time going from the past to the future. But this sequence has no meaning on the psychological level, for it unfolds completely without the knowledge of all the protagonists. What self s experiences, on the other hand, is a reversed process whose unfolding constitutes psychological time. There, self s declares itself the bearer and owner of desire d at nodal point N and desire d is scandalized to discover a desire D identical to itself and bearing on the same object, whose belatedness it will assert at nodal point N'. At nodal point N', then, desire d will assert

Figure 7.

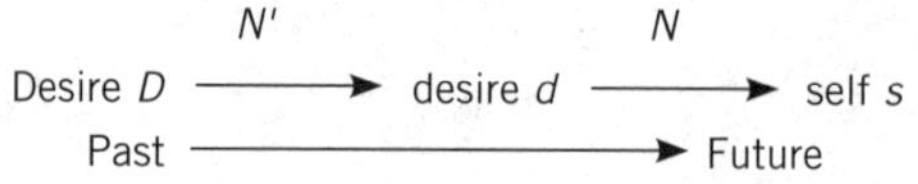

$$N' \qquad\qquad\qquad N$$

Desire D $\longrightarrow$ desire d $\longrightarrow$ self s

Past $\longrightarrow$ Future

its anteriority with respect to desire D. Such that self s, which is in reality the self-of-desire d, will lay claim loud and clear to the possession of the object of the two desires d and D.

This whole psychological sequence will constitute a new time—psychological time, the time of memory, the only time that has any meaning for the subjectivity of human beings, the only one that appears true and in accord with reality.

In *physical time*, that is to say in historical reality, the facts unfold as shown in figure 7. This physical time has no psychological reality although it is accessible to intelligence and therefore cognitive reality, but only when the first step toward wisdom is taken, that first step being the questioning or the recognition or the beginning of the recognition of the precedence of the other's desire, its priority over my desire, and therefore the nonownership of "my" desire. *Psychological time*, alone significant in lived experience, is the time of memory. It "climbs back up" physical time, as is shown in figure 8.

Let me open a parenthesis: the reason I say that physical time has no reality is that during a psychotherapy, when one revisits an event in memory, this event becomes present; psychological time can be made present but it is never correlated to physical time. For instance, I know that I have been in a certain city before, and I have vivid memories of some of the things I did or some of the people I saw in that city, and I can make those moments present to my mind now—but I am totally unable to tell you whether they occurred ten or twenty years ago. In other words, I am totally unable to evaluate the physical time between today and the time that my memory presents to me. "Time regained" in the Proustian sense is not physical time, because what Proust "regains" are not "durations" but sensations, emotions, and images.

Figure 8.

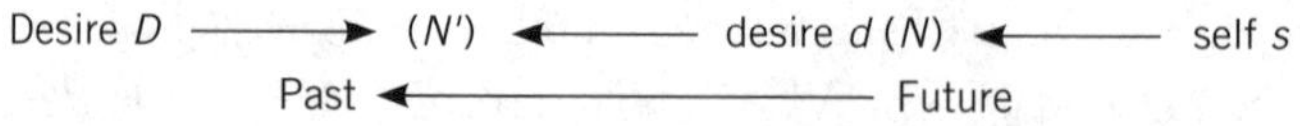

Desire D $\longrightarrow$ (N') $\longleftarrow$ desire d (N) $\longleftarrow$ self s

Past $\longleftarrow$ Future

One travels in psychological time in a way that is not possible in physical time. One can travel twenty or fifty years in a second, and suddenly remember an emotion, a sensation, a feeling or an entire scene that happened thirty years ago, and that scene becomes present before one's eyes. And that "presentification" has nothing to do with physical time because it is the resurrection of a moment that would take thirty years to recover if the recovery process had to be accomplished in physical time.

◆ ◆ ◆

If universal mimesis governs the human sciences in the same way that universal gravitation governs the physical sciences, the attraction exerted by a human being on another human being is proportional to its mass and inversely proportional to the square of the distance that separates them. The mass can be represented here either by volume (the attraction of an adult compared to a child) or by number (the attraction exerted by a crowd on a human being who is caught up in it).

But whereas our macroscopic bodies function according to these laws, our microscopic bodies (our atoms, photons, electrons, neutrinos, etc.) function according to the laws of quantum physics.

It seems to me that memory, too, obeys the laws of quantum physics and that further research in this direction could be enlightening.

But for the moment, I already see two very well-known mimetic phenomena that bear an obvious resemblance to the laws of quantum physics.

· A photon can be in two places at the same time. By the same token, in the world of memory, I can be at once in Los Angeles and in Paris.
· In the quantum world, the present and the future can influence the past. Insofar as I understand Schrödinger's theory, reality depends on observation and in the final analysis on consciousness itself. By the same token, in memory, no recollection is represented identically. Present consciousness reworks and so to speak re-creates memory. The latter is thus subjective and that is why the witnesses of an event never describe the same thing.

What makes for the specificity of points N and N' is that they are at once present and past, and the initiatory work consists in grasping them as past but continually present phenomena, in which capacity they maintain the self's existence. But we will see that the self and the desire that constitutes it have always and at every moment been at work to maintain psychological time in the quantum universe of memory. It is their action, in a certain sense, that continuously creates memory and maintains their existence. We will see that this existence can be maintained peacefully by a form of forgetting, or neurotically by a claim to the ownership of desire by the self or again psychotically by a frenetic reverberation of desire, affirming its precedence with respect to the other's desire that it has nonetheless imitated.

Psychology must therefore take account of the fact that the recognition of the otherness of "my" desire can occur at any moment and immediately modify all other misrecognitions. It is in this way that it can transform, that is to say have an initiatory effect, and lead to wisdom, a state in which time is unified, and becomes in a sense "nontemporal," that is to say always present, thus giving a glimpse of eternity.

At N, everywhere and always, in all neurotics, which is to say practically all humans, the self lays claim to the ownership of its desire, which historically (or in physical time) in fact gave birth to the self.

At N', desire d claims anteriority over desire D, which historically in fact engendered it, since d is mimetically inspired and produced by D.

It is however clear that nodal points N and N', which we are distinguishing here for didactic reasons, are in reality coextensive with each other and cannot exist without each other.

All of this leads to the conclusion that in psychology, physical time has no meaning and the future corresponds to nothing at all. Only the past related by memory is considered as real. And this past, this psychological time, is in reality the inverse of physical time, which means also the contrary of the reality of things. We will see that all the work of psychological initiation and of the quest for wisdom consists among the great initiates in realizing that what their memory reports is exactly the reverse of what actually happened.

The realization of this fact is what I call the *recognition* of the otherness of my desire, which leads to peace and wisdom. Conversely, the *misrecognition* of this otherness—which is expressed by the frenetic claim at N of the ownership of "my" desire and at N' of its anteriority with respect to the other's desire—leads to all psychopathological syndromes, for both neurotics and psychotics generate diverse and multiple strategies to make good on this double claim $(N—N')$: the desire I feel belongs to me and it has priority over the other's desire that appears to be copying it.

It is thus in psychological time—the only time that has meaning—that the issue of recognition or misrecognition can be decided. It is also to be stressed yet again that *recognition and misrecognition will take place in every single minute of physical time and that recognition, if it is to happen, will happen physically "now," in the present, but will enlighten and heal all the preceding, earlier misrecognitions.* Here we return to Proust's insights about the search for lost time and time regained. As for the future, on the psychological level it represents nothing but a projection of the past and an imaginary repetition of memories.

Everyone thinks that memory is a machine for storing memories and thus a recollection apparatus. But it is clear that for self s to be able to remain in existence, it must forget the problems raised by nodal points N and N'. Memory thus has an essential function, which is the forgetting of the genesis of desire and of the self, enabling the latter to remain in existence with all its attributes. The proof is that when—during a hypnotic trance, when the vectors are reversed once again—the hypnotizer's desire D creates desire d', which in turn generates self s' with all its attributes, self s dissolves and disappears.

Four Kinds of Imitation

I distinguished between four forms of mimesis bearing successively on appearance, belongings, being, and desire.

· As for *appearance*, imitation is Platonic: it is a matter of imitating the form, and the form that potentially contains all others is the "idea." Thus, the idea of a table contains potentially all tables that can be manufactured; and in turn, a particular table contains potentially all the representations that can be made of it in painting, photography, and so on. This leads us to a fundamental notion that perhaps escaped Plato, namely that imitation can add or subtract information from the imitated model. Thus, when Da Vinci paints the *Mona Lisa*, he loses an enormous quantity of information concerning her: her odor, her voice,

Mona Lisa in tears, Mona Lisa laughing, and so forth; but he adds an enormous quantity of information to the real young woman in front of him, to such a degree that the copy he makes of her will live on through the centuries. The same holds true for the imitators and caricaturists who, by exaggerating, add information to the simple observation of a model while also losing a large quantity of data.

· When mimesis bears on belongings, the imitated gesture is a gesture of appropriation and quite simply of ownership. Wanting to appropriate what the other has is obviously a source of conflict and escalation that can be very violent. This is what Cesáreo Bandera calls *mimesis conflictiva*.[1] It is what Girard describes as "appropriative mimesis," and it leads directly to violence. Of course, Girard tells us that appropriating somebody else's belonging is always transitive to acquiring his being. That is a violent way of achieving or pursuing being, except in advertising, where we can acquire some of the being of George Clooney by drinking Nespresso.

· Whereas appropriative mimesis engenders violence in most instances, mimesis that bears on the model's very *being*, amounting to an identification with this model, acts as a mechanism of appeasement and conflict avoidance because it suppresses the need to appropriate the other's belongings in order to reach his being. Indeed, if a child identifies with its father, this does not necessitate the death of the father except perhaps on the symbolic level, nor does it necessitate the appropriation of the father's tie, suit, shoes, car, or wife. So identifying with one's father does not entail any conflict or violence with him. We have here to bear in mind the dialectic between the father and the son, in other words between the model and the imitator. Because if the father, for instance, abuses the child or is violent and beats him up, the identification becomes very difficult in the moment but will resurface later when the child will be on the street attacking other adolescents or robbing banks. And this takes us back to the origins of the Oedipus story because when Oedipus meets his father on the road, the father starts by whipping him, insulting him, and telling him to get out of the way, which leads Oedipus to kill him. But obviously if the father had said to him, "Young man, I am old enough to be your father, so please let me pass," the rest of the story would not have happened. In a certain manner, mimesis bearing on the model's very being can often appear as a form of therapy or consolation, for the model's possessions are considered as his legitimate property.

· Finally, mimesis can bear on the other's *desire*, causing an identical desire to bloom in the disciple. Here, the adversarial or nonadversarial evolution of the mimetic desire will depend on the greater or lesser proximity to the model (as we saw above, when Girard makes the distinction between an "external mediator" and an "internal mediator"), but also on the nature of the desired object. If, for example, the desired object is literary glory, this can lead only to the production on the model's and on the disciple's part of more and more brilliant texts. But, here again, we will see that things can go wrong in the sense that mimetic desire can evolve toward the production of psychopathological symptoms.

At this stage in our reflections, we see that the first psychological movement, whatever its future complexity, always comes from the other and from the relationship with the other.

Myself Is an Other

I would like to emphasize here once again the notion of otherness. The desire that constitutes my self is, as we have seen, the other's desire. This otherness with which we are saturated and that constitutes us is the human condition; but it is very difficult to accept. Its misrecognition is initial and necessary to the maintenance of the self in its existence. Recognition is a difficult, initiatory enterprise, strewn with obstacles, and it is the key to mental health, to happiness and wisdom—we will come back to this. Partial or total misrecognition, for its part, can be peaceful in the form of "forgetting" or neurotic and frenetic at N, or again delusional and psychotic at N'. It is to these figures of psychopathology that we will return in the next part.

Let us come back to the Garden of Eden for a moment. God breathes his desire into the clay and creates man. The latter is thus entirely drenched in God's otherness, and that is why he is created, the Bible says, "in the image of God." God inscribes in him, from the instant of creation, the spatial dimension of universal mimesis, and of course, as we said above, there is a prodigious loss of information. Then God creates woman from a rib or rather from a "side" of Adam: woman is thus also entirely made up of otherness.

It seems to me that the misery of the human condition lies in the difficulty of accepting the otherness of one's own being, of accepting that myself is an "other" and that this other who constitutes me is anterior to me. The tragedy of the human comes from denying having been created by the desire, the breath of God, out of nothing (or almost: dust) and that one is permanently re-created by the desire of the other at each instant of one's life. The history of humanity and the history of each one of us is that of our revolts against the recognition of this otherness, that of our claim to originality, our anteriority, and of the priority of our desire over the other's desire, which inspired, induced, elicited, and created it. In other words, our story is tragic because it is the long string of sterile attempts to deny the real and escape it. Let us also recall the following: desire, which is mimetic, is radically distinguished from need, from instinct, and from drives and is revealed to be capable of perverting, subverting, or even suppressing them, as we shall see.

Desire d, the imitation of the other's desire (desire D) is from the first instant and by definition in the process of laying claim to its own priority. Self s, forged by desire d, must, in order to maintain itself, from the moment of its emergence, claim priority for this desire of which it is in reality the product. From this comes the fact that the geneses of d and of s fatally involve an uncompromising claim and thus a dose of rivalry. Thus, desire and rivalry are but one. There is no desire without rivalry. There is no rivalry without desire.

In the student or the apprentice, desire is not very rivalrous. The misrecognition of otherness—which is initially necessary, let us recall, to keeping the self in existence—is peaceful and is manifested as an overlooking or bypassing of the problem. The teacher teaches and the pupil learns. Everything goes smoothly until the day when the student feels that he has surpassed the master. Then rivalry can emerge at N with neurotic frenzy and at N' with delusional frenzy.

To sum up, human beings are the plaything of mimetic mechanisms, of rivalries that emerge out of imitations. All of this occurs at the level of the interdividual rapport, which in the following chapter I will propose we individualize under the name "third brain." These destructive, demanding, rival mechanisms express the refusal to recognize otherness, clothe themselves in sentiments, in fury or coldness and various emotions taken from the emotional or second brain's wardrobe. And they don political, religious, philosophical, ethical, and other justifications and rationalizations taken from the wardrobe constituted by the cognitive brain, which very soon we shall label as the "first" brain.

These basic mechanisms are camouflaged beneath the mythologies, allegories, and dazzling lucubrations of philosophies and psychologies through the ages: culture, by all possible means, attempts to protect us from the real, from the recognition of our otherness; culture is fully complicit in our misrecognition and always ready to excuse our protests and our revolts. Greek mythology, which transforms human rivalries into the combats of the gods on Mount Olympus, like psychoanalytic mythology, only masks, disguises, and prettifies the elementary, banal, rather unattractive and repetitive mechanisms that manipulate us. That is why the reader will have the impression in this book that I am always saying the same thing, because I am inventing nothing.

The Three Brains

The First Brain: Cognitive and Rational

The ancients did not attach great importance to the brain. In ancient Egypt, the heart was considered the seat of perception, cognition, and the soul. That is why, during the mummification of the pharaoh, his brain was removed via the nose and discarded—the organ could be of no use to the deceased in the afterlife!—whereas the heart was carefully preserved.

Aristotle himself thought that the heart governed cognition and perception. In his system, the brain's function was to cool the heart's passions. In this sense, he was already close to a conception where the emotional apparatus that generated the passions was tempered and "refrigerated" by an apparatus that was calming and . . . reasonable.

Leonardo da Vinci thought that perception and cognition resided in the ventricular cavities of the brain and not in the cerebral substance itself. These functions were nevertheless already located in the head.

For Descartes, the soul and the body were of a different nature and essence. He thought that their point of interaction was located in the pineal gland, the only asymmetrical structure in the brain. Thought, for him, could not be produced by brain matter; it issued from the rational soul and had an ontological, spiritual, and metaphysical character.

Beginning in the eighteenth century, doctors and scientists began increasingly to consider the brain as the seat of the function that up till then had been attributed to the soul. In 1687, Thomas Willis proposed the term "neurology" to qualify the discipline charged with studying brain illnesses. Willis was one of the leaders of the "Oxford group," which included doctors and philosophers like Robert Boyle, Robert Hooke, John Bock, and Christopher Wren. In *The Anatomy of the Brain and Nerves*, published in 1664, he maintained that perception, movement, cognition, and memory were functions of the brain itself. From this moment forward, the brain became the object of study of neurologists and psychiatrists, but the functions studied were essentially cognitive functions: reason, judgment, the five senses, memory, and motor functioning.

Ever since, psychiatrists and psychologists have never spoken of anything but the first brain. And the latter is the only one to have answered them. It has always been their only interlocutor. Apart from Spinoza who, as we have seen, gave to "affect" a primordial role in determining behavior, the philosophical tradition as well has always involved the solitude and independence of the brain thinking and reflecting in the secret of a study and mastering the physical and metaphysical world by means of thought. "I think, therefore I am," said Descartes, thereby locating being in the brain or at least in the part of the brain I am now calling the "first brain."

Not only does this first brain enable us to move, walk, use our five senses, see, understand, and so on, but above all it allows me to write this book and permits you to read it. This first brain is thus of capital importance, but it is not unique and it does not work all alone.

For a long time, then, it was believed that the human being had but one brain. The cortex was mapped and divided into motor areas and sensitive areas to which were added the sensorial areas, the receptacles of the five senses. Language was located in the left temporal regions (Broca's and Wernicke's areas).

The first brain, or cognitive brain, was thus fundamentally the seat of intelligence, which psychologists measured by means of IQ, or intelligence quotient; it was thus the seat of cognition and rationality, of the comprehension of things and of the world, and thereby of rational exchanges and relations among rational people. Finally, this brain was the exclusive seat of memory, which could be strong or weak, its "flaw" being forgetting. It was Freud who established the distinction between conscious and unconscious memory, the unconscious being constituted not by mere instances of forgetting but by the active repression of traumatic memories, which could not be summoned by consciousness but played a role in every human being's life and destiny.

Plato knew that he thought with his brain and not with his stomach or legs. But the doctors of Greek antiquity already suggested the possibility of another sort of brain, that is to say of another self endowed with its own life, thus relativizing the independence and power of rational thought. The problem was to identify and localize this "counterpower." It was, for the ancient Greeks (notably Hippocrates and Galen), left to the sexual organs and especially the female sexual organs to explain hysteria; then for centuries it was the demon that led astray or possessed minds, and finally, it was the Freudian unconscious, whose great merit is to have situated the counterpower in the very heart of the brain and in particular of the first brain. We will come back to this.

But before we do, let us examine in greater detail another discovery in neuroscience, neurology, and psychology that emerged at the end of the twentieth century: the discovery of emotional intelligence. Beginning with the research of Antonio Damasio, most notably, the existence and importance of a second brain was brought to light: the limbic brain.

The Second Brain: Emotional and Affective

The study of mirror neurons—the discovery of which came scarcely ten years after that of the limbic brain—has taught us a great many things, in particular that the mirror system could well offer an explanation for empathy, as Pierre Bustany, professor at Caen, underlines at each lecture he delivers. Empathy has taken on considerable importance to the point that the American thinker Jeremy Rifkin, in a recent essay, *The Empathic Civilization: The Race to Global Consciousness in a World in Crisis*, speaks of *Homo empathicus*. In an interview with the French magazine *Le Nouvel Observateur*, Rifkin criticized the philosophical tradition for refusing to consider the empathic dimension of human nature. "Recent research in the cognitive sciences show this: babies are empathic beings, sensitive to the suffering of others. This is what makes us human beings, it is our specificity. Bizarrely, however, the notion of empathy has never interested thinkers."[1] He notes that for Hobbes human life is solitary and indigent, and that man is "fundamentally aggressive and wicked." Locke, according to Rifkin, is less pessimistic, but thinks that our ultimate mission is to be productive beings. Adam Smith sees us as beings guided by greed and the maximization of profit. As for Jeremy Bentham, "For him, the human condition is reduced to avoiding pain and seeking out pleasure." Rifkin suggests that this vision was later taken up by Freud with his pleasure principle and death drive. He asks: "But what if, to the contrary, as the latest discoveries in neuroscience suggest, man is first and foremost a social animal?"[2]

I think Rifkin caricatures the thought of the philosophers he mentions, but what is important is that on the one hand he notes what I myself wrote in 1982, namely that psychology and sociology are indissoluble and form a single science, and on the other hand he accords empathy, in the wake of the discovery of mirror neurons, tremendous importance. He underlines the relevance of the research that establishes the existence and the crucial role of the limbic brain because of the likely presence in the second brain, as in the first, of mirror neurons and of a mirror system that may well explain emotional and social intelligence.

The discovery of the second brain can be credited to numerous researchers in neurology and neuroscience, first and foremost among them Joseph Le Doux and Antonio Damasio. Emotional and social intelligence were then popularized and spread among the general public through the writings of Daniel Goleman. In *Emotional Intelligence*, the latter writes: "We have two minds, one that thinks and the other that feels."[3] For Goleman, the emotional and rational minds are "semi-independent" faculties. Each reflects "the operation of distinct but interconnected neural circuits."[4] The two brains are wonderfully coordinated, writes Goleman. Feelings are as essential to thoughts as are thoughts to feelings, but when passions flare up, the balance is thrown off: the emotional brain takes control, sweeping the rational brain aside.[5]

It was the study of a very unusual patient, Phineas Gage, wounded in 1848 on the construction site where he was working, that led Damasio to the discovery of the emotional brain: "Entering from the left cheek upward into the skull, the iron [bar] broke through the back of the left orbital cavity (eye socket) located immediately above. Continuing upward it must have penetrated the front part of the brain close to the midline."[6] In surprising fashion, Phineas Gage recovered in less than two months, but his "disposition, his likes and dislikes, his dreams and aspirations, all would change."[7] Gage's body was still alive, but another personality now seemed to inhabit it. Gage was no longer capable of making a decision or of conducting himself normally in society. We must thus form the hypothesis, Damasio tells us, that "normal social conduct required a particular corresponding brain region."[8]

The second brain is constituted by the limbic system, which essentially includes the prefrontal cortex and in particular the ventromedial region of the frontal lobe, the hypothalamus, the gyrus, the amygdala, and the structures at the base of the telencephalon. The amygdala, notably, seems to be in charge of emotional memory and thus of the affective and subjective meaning of things. Goleman recalls that if the amygdala is disabled, the subject becomes affectively blind, incapable of making emotional sense of events. The work of Joseph LeDoux has shed particular light on the role of the amygdala, underlining that its interaction with the neocortex is at the foundation of emotional intelligence. And Goleman writes that the brain has two mnesiac systems, one for ordinary facts and another for facts laden with emotion.

It seems thus that there is a cognitive and intelligent memory in the first brain and an emotive and affective memory (the one that Proust was particularly interested in) in the second brain. Daniel Goleman tells us that emotional intelligence includes "abilities such as being able to motivate oneself and persist in the face of frustrations; to control impulse and delay gratification; to regulate one's moods and keep distress from swamping the ability to think; to empathize, and to hope."[9]

It is thus clear that the rational neocortex, in which the sensitive, sensory, and motor function zones are located, and which, as we have seen, contains a mirror system, is not enough to ensure the harmonious functioning of the psychological organism. There must be permanent interaction with the second brain, the seat of emotions (joy, surprise, fear, anger, disgust, and so forth) and feelings (love, hate, resentment, envy, jealousy, and so forth) but also of mood (good or bad, excited or depressed, accelerated or slow). This second brain, let us recall, also seems to be equipped with a mirror system explaining empathy and the comprehension, transmission, contagion, and sharing of feelings, emotions, and mood.

The Third Brain: Mimetic and Relational

As I have said, starting in 1978, René Girard, Guy Lefort, and I thought that there is no psychology except for interdividual psychology. An infant who has just come into the world relates to others in a way that is first and foremost and only mimetic, and this includes not only the motor aspect (the first brain) but also the affective aspect (second brain). In 1982 I also put forward the hypothesis that mimetic desire—and more generally the mimetic mechanisms bearing first on the model's appearance, then on the model's belongings, then on its being and finally on its desire—is at the origin of the birth and existence of what, in each one of us, could be designated as the "self."

By a completely different path, Antonio Damasio has arrived at views identical to mine. He speaks of the necessary neural foundations of the self and adds that the self is "a perpetually re-created neurobiological state,"[10] noting later on: "There would be successive organism states, each neurally represented anew, in multiple concerted maps, moment by moment, and each anchoring the self that exists at any one moment."[11] Finally: "At each moment the state of self is constructed, from the ground up. It is an evanescent reference state, so continuously and consistently *re*constructed that the owner never knows it is being *re*made unless something goes wrong with the remaking."[12]

These views totally corroborate my metapsychological theory of the desire-self outlined above, as well as my reading of the hypnotic phenomenon as laid out in chapter 5 of *The Puppet of Desire*. I would add to Damasio's remarks that the self is continually *shaped by the mimetic mechanism within the interdividual rapport*.

With respect to the self-of-desire, a friend of mine who is a psychoanalyst wrote to me: "And yet there exists a self that is, to be sure, modifiable *in part* by encounters and interactions, but which *preexists* and is more or less solid." Let's take a look at Freud's second topic to explore this idea further. In the formation of the self, Freud emphasizes the central and fundamental role of the child's relationship with the mother. If we wanted to express this in our vocabulary, we would say that the interdividual rapport with the mother fashions and nourishes the self-of-desire, which is thus at first the self-of-the-mother's-desire. This desire must therefore be the strongest, the most positive, and the most ambitious desire possible for the child. But this desire would be vain if it was not accompanied by the mother's love, which, in our vocabulary, constitutes the affective reserves of the second brain. This maternal love will in some sense be "stored" in the second brain to be dispensed throughout the child's life: it will be a source of positive feelings like the capacity for loving, and positive emotions; in all circumstances it will be anxiety-decreasing and reassuring, removing fear.

The "Freudian" father—that is to say the father at the end of the nineteenth century—represents taboo, and the identification with the latter, that is to say the imitative integration of that taboo, is baptized "superego." This prohibition is loving and benevolent, and not categorical: the father establishes laws—"We eat when we're at the table, you can play later," "You can have a bicycle when you're older," "We'll go to the countryside tomorrow if you behave." In other words, this superego reinforces desire through taboo in that it teaches desire to delay gratification and thus to maintain itself as it awaits its realization. Paternal taboo, the superego, teaches desire "differance" (in the Derridean sense), that is to say teaches it to survive beyond the present moment, the impulse, to strengthen itself, to endure, in other words to transform itself into willpower.

As for the Freudian "id," it provides desire with the energy necessary for its realization. I have addressed this subject elsewhere[13] and affirmed that desire, being mimetic, acquires its energy from its rapport with the other— we shall have examples of this later in this book.

The Freudian self is thus from the very beginning saturated with otherness. The mother's and father's role in the constitution of what could be called this "first self" is fundamental. But this self will complete itself gradually via all the imitated desires and all the new selves that are created. The self will be perpetually reshaped, made up of a patchwork of all the selves formed in the course of its history.

That being said, this first self, the self formed by the parents, is perhaps the one that my psychoanalyst friend is talking about. I would be in agreement if there were many such selves left: in the twenty-first century, how many mothers still have for their children the absolute love described by Freud? How many fathers still have authority and are capable of laying down the law / establishing boundaries? If we could still find first selves formed this way, I would be willing to concede that they would be more "solid" than the others. Perhaps, too, selves of this sort are more capable of "resiliency."[14]

To come back to mirror neurons: they have thus taught us that these mimetic mechanisms initiate the action of the two brains; they constitute the system by means of which humans enter into relationships with one another. The mirror system of A is thus the first and the only to enter into a relationship with B, and that is how what I have called the "interdividual rapport" is established and created.

I think the time has now come to grant this mimetic interdividual rapport (which in my opinion is likely based upon the activity of the mirror neuron system) all its importance by distinguishing it as the third brain. This will lead us to construct a new anthropology and to understand psychological and psychopathological constitution as resulting from the interactions and the balance among the three brains.

It seems interesting to me here to cite Jacques Lacan, who writes in the introduction to his doctoral thesis in medicine, *De la psychose paranoïaque dans ses rapports avec la personnalité*: "There exist mental disorders that, attributed, according to one's doctrine, to "affectivity," to "judgment," to "conduct," are all specific disorders of psychological synthesis. . . . We call this synthesis *personality*."[15]

I think that at this stage the imbrication of the mirror system with the first and second brains is clearly established, as is the indispensable coordination between the first and second brains. Lacan evokes the second brain when he speaks of "affectivity," the first when he speaks of "judgment," and, it seems to me, the third when he speaks of "conduct." He says that mental disorders are due to a lack of "synthesis" among these three elements. This is evocative of the most recent discoveries in neuroscience, neurology, and psychology in the several decades since Lacan wrote his thesis in 1932.

While it was thought to be sufficient, in order to account for the function of the psychic apparatus, to study the necessary and complementary relations between the cognitive or first brain and the emotional or second brain, it appears to me indispensable to assert that the third brain, the mimetic brain, is the one that introduces the little human to sociability, to relations with others, to the interdividual rapport, and indeed to humanness.

The Three Possibilities
of the Interdividual Rapport

I t is at the level of the interdividual rapport that mimetic desire is born. The latter draws its energy as well as its aim, that is to say the choice of its object, from the imitation of the other's desire. This imitation, as we have seen, can entail learning and progress or, to the contrary, rivalry, conflict, violence.

Model, Rival, or Obstacle

Broadly speaking, the relation to the other can be broken down into three possibilities: the other as *model*; the other as *rival*; the other as *obstacle*.[1] Within this framework, we will see that there are very fine gradations, that all of these variations must be taken into account, placing emphasis on continuity rather than on discontinuity. In his preface to *Psychologie de la vie amoureuse*,[2] Robert Neuburger illuminates the similarity between Freud's vision and my own with respect to patients: "There are not two human species, healthy people and sick people. Pathology sheds light on normality; there is a continuum between attitudes that are apparently far from the norm and what can be expected from allegedly normal people."[3]

When desire takes the other as a model without rivalry, we find ourselves in the framework of learning and friendship, that is to say in a situation of mutual, continuous, peaceful imitation-suggestion. This explains why friends often have the same tastes, like to go to the same places, listen to the same kind of music, and in general are "on the same wavelength." Each serves as the other's model and each imitates the other's suggestion as well as possible.

Learning, too, implies the spontaneous participation of the student, but also of the teacher. For example, if I want to learn to fish, I can hide behind a tree and watch what the fisherman is doing, but what I am doing is really nothing more than spying and I am unlikely to make any progress. On the other hand, if I accompany an experienced fisherman and he shows me how to make the right motions and I imitate them under his supervision, I am a good student and I learn more effectively. For there to be learning properly speaking, the model must behave as such and deliberately and consciously try to pass his knowledge or technique on to the student while helping the imitator to improve. The latter must want to learn, and the model, the teacher, must want him to imitate.

When I take the model as such, this relationship resonates in the first brain as politically and morally justified, economically valid, or religiously recommended. In the limbic system, this relationship elicits positive emotions such as admiration and love for the teacher, as well as a good, stable mood. When the other appears as a rival, on the other hand, appropriative mimesis leads to my tearing away the object that he pointed out to me. The interdividual suggestion-imitation rapport experiences an escalating process of mimetic rivalry due to our common desire.

The model has become a rival! This is the kind of situation that leads one to come to a psychiatrist's office. That is why I have repeatedly said that the clinical expression of mimetic desire is rivalry. As I already remarked in Part 1 of this book, never would a patient (let's call him David) come to my psychiatrist's office and tell me: "I have a problem, I'm imitating my friend Pierre." He would instead come and say to me: "Listen, I have a problem with my best friend who I am realizing is in fact a bastard who is trying to take away the girl of my dreams." His forgetting bears on the fact that his desire is mimetic—and thus identical—to Pierre's, and this forgetting leads the patient to affirm the *ownership* and at the same time the *anteriority* of his desire with respect to his friend's. From the patient's point of view, it is Pierre who is imitating him rather than the other way around! David is seeking to establish fallacious differences between his desire and his rival's. When rivalry sets in, the third brain is readily fed by intellectual and moral justifications stemming from the first brain. In other words, from David's point of view his friend-enemy Pierre is wicked and ill-intentioned.

The third possibility is for the relation to evolve, turning the model into an obstacle. In that case, the disciple's desire to acquire the model's belongings or being will appear radically impossible. Suppose Pierre succeeds in taking away David's girlfriend and David desperately tries to recapture her until the girl herself tells him she doesn't want to see him anymore: "I love Pierre and I'm afraid you don't have a chance with me." The erstwhile friend then becomes in David's eyes an insurmountable obstacle, because the very object of desire has chosen the victor. And we shall see in clinical practice the various aspects that this situation might entail.

◆　◆　◆

Once again, I want to attract attention to the fact that the first and second brains are, so to speak, pulled along by the third brain. According to the relation's tonality, then, according to whether the other is perceived as a model, rival, or obstacle, the quality of the interdividual rapport, that is to say of the mimetic relation, reverberates its effects on the cortical and limbic brains: it will go into the wardrobe of the first brain in order to coif itself in economic, political, moral, or religious justifications and rationalizations, and in the wardrobe of the second brain to dress itself in the matching emotions, feelings, and moods.

Now looking at things from the point of view of the third brain, that is to say analyzing the quality of the relationship to the other and the nature of the interdividual rapport, is not without implications: the result is to upend the whole field of psychopathology. Mirror systems establish between one another primordial rapports whose evolution unfurls, in perfect continuity, the various figures of psychology and psychopathology.

"How" Instead of "Why"

At this stage, a fundamental, and I would almost say ontological, problem is the following: what is the role of each person's individual constitution in interdividuality? This problem brings us back in a certain way to the debate initiated earlier by my psychoanalyst friend. Indeed, if interdividuality determines the psychological and psychopathological destiny of human beings, those human beings nonetheless enter into mimetic relationships with a particular history and structure. To illustrate this, I will propose the following example: picture a lecture by Krishnamurti in Saanen, in Switzerland. Krishnamurti appears as a sage whose teaching is based on a certain number of recurrent notions and who does not wish to keep for himself knowledge and wisdom but to the contrary wishes to share them with the greatest number of people possible. He is thus objectively an open model.

A portion of those who are listening to the lecture are enthusiastic, enchanted to rediscover familiar themes developed in another way, and to understand them better. These listeners find that they are encouraged in their quest for a certain kind of wisdom, and their admiration for Krishnamurti is colored by the second brain with affection and gratitude. For them, Krishnamurti is a model and he remains thus. Their first brain is nourished by his teaching, the second has only positive and warm feelings for him, and the third brain is fixed in the "model" position.

A second category of listeners gets irritated and annoyed and tell themselves that after all Krishnamurti keeps saying the same thing, that he's not bringing anything new to the table, and they don't see how they can adopt his way of seeing things or learn to live any better thanks to him. Their first brain does not follow along, the second brain furnishes negative feelings of impatience and annoyance, not to say antipathy, toward the lecturer, and their third brain causes the interdividual cursor to get stuck in the "rival" position.

A third category of listeners in the audience is overwhelmed by the lecturer and considers that what he is saying is so wise and marvelous that they will never manage to understand and still less to imitate him or apply the formulas of wisdom that he is offering. The latter will sink into discouragement, and a slight depression will alter their attention to the point that soon they will be unable to follow Krishnamurti's words. This renunciation bears witness to the fact that the cursor of their interdividual rapport is stuck in the "obstacle" position. Their first brain is filled with admiration but convinced of its incapacity to understand and to follow, and the second brain produces sad feelings of discouragement and abandonment.

Later on, these three categories of listeners will be dispersed, but for a reason that is difficult to grasp, they will continue to have their interdividual cursor stuck on the preceding position in all their human relationships and in all the situations that they encounter. This is a very important point, for it completes our understanding of "Girardian" mimetic desire by showing that rivalry, for example, is not always the result of a model that is trying to keep an object for itself, and that the model-obstacle is not always the one who forbids possession of the object in a radical way. The debate where this point is concerned could consist in wondering what causes the mimetic cursor of the interdividual rapport to become fixed in the "model," "rival," or "obstacle" position. This etiological search could be interesting, but it seems to me to lead to an impasse. The predisposition of each to more or less rivalry is a fact that must be accepted, and from there it is better to describe the phenomena that it engenders. Relational psychoanalysis might be able to explain what in the past of those three categories of people explains or determines their present attitude taking the model as a model, a rival, or an obstacle by the reproduction of past patterns. Yet after saying that relational psychoanalysis could possibly explain mimetic predispositions by the reproduction of a former pattern, the question remains as to whether this pattern can ever be changed. As the reader will see at the end of this book, I have a suggestion as to how to go about such change.

From my point of view, the question that has meaning in psychology and psychopathology is "how": how psychology works and how it can be modified for the sake of improvement. The question "why" must be avoided, first of all because it immediately runs up against binary oppositions—nature versus nurture, the psyche versus the body, and so on—and then because it often ends up falling into the trap of the scapegoat: we look for a cause and we find . . . a culprit. In each case, as we've seen, the interdividual rapport will seek in the wardrobe of emotions and feelings that makes up the limbic brain something to dress up the triggered mimetic mechanism; then it will seek moral, ethical, logical, philosophical, and rational justifications for the mimetically induced action.

Figure 9.

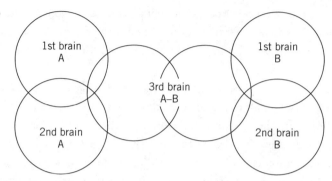

At this stage, we can depict in a very simple diagram the architecture of the psychological apparatus as we have been led to conceive of it (figure 9). Is there any need to specify that the three brains that we are separating for didactic purposes are in reality interconnected by billions of synapses and that the whole forms an indivisible unity?

What about Collective Intelligence?

Finally, let us remark that the discussion about collective intelligence can be enriched by the contribution of mimetic theory. How can it be that the intelligence of two or several human beings, grouped into a team (as opposed to a large group or a crowd), is superior to the sum total of the individual intelligences that make up the group? What happens between brains A and B, thinking over the same problem, for example, is the immediate constitution between them, through the play of empathy, of a third brain AB that they have in common. The harmony and cohesion that reign in the interdividual rapport multiply the energy of third brain AB, instead of leading to the mere addition of first brains A and B. This harmony in A and B extends to their first and second brains, mobilizing all the cognitive and imaginative capacities in their first brains, and all the positive, affective, and emotional energy in their second brains, which moreover colors this cooperation with a joyous harmony.

Thus the harmony and peace between A and B and among the three brains of A and B is more powerful, richer, and more productive than the rivalries, conflicts, and inhibitions that emerge if, in the interdividual rapport, A and B are rivals or obstacles for each other: two or several third brains in empathy and peace, in collaboration with one another, increase their power and that of their first and second brains.

But there is more: the interdividual rapport, tightened and strengthened between two or several actors, creates in each of its poles what could be called a "self." This is the emergence of a new "self" that will itself be surprised by its new imagination and its new intelligence, which its usual "self" couldn't even have predicted. It is these new "selves," dual or collective, generated by the otherness of desires united in a single enterprise, that will produce the "miraculous" discoveries of collective intelligence. (Of course, we have to underline the fact that beyond a certain number, when the team grows into a large group or a crowd, then the mimetic entrainment might lead to the most ferocious or irrational behaviors.)

The third brain is a platform of mimetic exchanges, engendered by the mirror system and made up of thousands of imitations and suggestions ceaselessly circulating between A and B. Of course, A and B are also engaged in relations with C, D, E, and so forth, that is to say with thousands of other entities, both material and cultural, which make of each of us patchworks of all the influences we undergo, all the brains we encounter.

CHAPTER 8

Classical Nosology

U p to the present in psychopathology and psychiatry, mental disorders were treated based on the idea that in psychoses the cognitive brain is malfunctioning and in neuroses the emotional brain is malfunctioning. Psychiatrists did this without realizing it, but in light of all the recent discoveries that we have just examined, this fact will appear clearly. A brief overview of classical European nosology seems to me in order before we go any further.

Psychoses

Psychoses are characterized by delusion, in other words by a discourse about all or part of reality that is delusional. Classical European psychiatry characterizes delusions by three elements: their mechanism(s), their theme(s), and their structure(s)—the "structure" referring to the fact that they are "systematized" or "nonsystematized" (i.e., they possess—or do not possess—an internal coherence). These three elements obviously fall under the cognitive brain, the neocortex or first brain. Four psychoses are individualized:

1. *Paranoiac psychoses* are characterized by a single theme: demands, persecution, invention, or erotomania (the delusional conviction that such and such a famous person or star is in love with me), or else a hypochondriac complaint; a single mechanism: interpretation, sometimes (and even often) based on or completed by a secondary mechanism, intuition (if a paranoiac goes out of his house and sees a person parked in a car, he will have the *intuition* that this is a spy and *interpret* the presence of the car and the person in it as a proof that he is being spied upon); a perfectly systematized structure, that is to say having internal coherence and sounding like a rational story.

2. *Chronic hallucinatory psychoses* (CHP)[1] are characterized in the same way by a single theme, persecution, or sometimes an erotic-mystical theme; a single mechanism, hallucinations that can be auditory most of the time but sometimes also olfactory and even cenesthetic; a relatively well-systematized structure. It is with respect to these CHP that de Clérambault described "mental automatism" or the feeling of having one's thoughts stolen, wherein the patient is convinced that others (individuals or government organizations) can read his thoughts and impose theirs, often while offering commentary on his actions and giving orders in a clear voice that he is the only one to hear!

3. *Schizophrenias*, setting aside all that can be said about discordant cognitive function and ambivalent affective function, are diagnosed based on three simple criteria: multiple themes, bringing together persecution, mysticism, erotic or fantastic themes, and so forth; multiple mechanisms, at once hallucinatory, interpretative, imaginative, intuitive, and also mental automatism and disordered train of thought; a totally nonsystematized structure, in other words a completely incoherent discourse.

4. *Paraphrenias* are rare psychoses because they are rarely diagnosed because they often remain secret, "encysted," and for the sole use of the deluded person; here again, the diagnosis rests on a single theme, which is fantastic or more often related to filiation; a single mechanism, the imagination; a relatively well systematized structure.

Thus diagnoses of psychoses in classical psychiatry are made solely based on the analysis of discourse, that is to say of the patient's *delusion*, which translates or expresses the disorder of his cognitive brain. That is why these psychoses are called "mental illnesses."

Neuroses

Classical psychiatry recognizes four neuroses:

1. *Obsessional neurosis* in its different varieties—compulsive cleanliness, compulsive hoarding, compulsive counting or repetition—is also a mental illness, because although consciousness is clear and unaltered and reasoning logical, a mysterious force obliges these patients to wash their hands twenty times in a row, to clean their room ten times in a row, to check the locks dozens of times, to accumulate the most varied objects because of an inability to throw anything out, and finally to practice often complex rituals (like counting one's footsteps, for example). The failure to carry out a ritual or a compulsion leads to great anxiety. In these neurotics, what seems to be malfunctioning is essentially cognitive in nature: knowing that what one is doing is absurd, but being unable to prevent oneself from doing it. It seems very plausible that all of this unfolds in the first brain, but if the latter misses a step in the ritual, the second brain goes into panic mode and brings the patient to seek psychiatric help.

2. *Anxiety neurosis* falls squarely under a disorder of the emotional brain. Against a backdrop of permanent anxiety, paroxysmal attacks of anxiety are triggered, without logical or rational justification, thus apparently without cortical participation—and they are therefore incomprehensible. These attacks essentially present somatic manifestations: heart palpitations, sweating, suffocation/breathing problems (hyperventilation), stiffness/rigidity, feelings of pins and needles in the hands and feet, and so on. These attacks have been baptized "tetany," "spasmophilia," or "panic attacks," and their etiology has been ascribed to the most varied causes: magnesium shortages,

hyperventilation, premenstrual hormonal problems, and so on. In reality, it is clear at present for us that anxiety, whatever its origin may be, is expressed by a storm in the limbic system whose role as biological and hormonal "conductor" (in the musical sense) explains the multiple and varied somatic manifestations observed.

3. *Phobic neurosis*: if anxiety is fear "of nothing," that is to say nothing definite, fear without an object, phobia looks like a mechanism for mastering anxiety. The latter, in the phobic neurosis, is focused and localized in an object, a situation, an idea. It then "suffices" to avoid that object, that situation, or to fight against that idea, to diminish anxiety. For example, if you "decide" that you are afraid of mice, it will be sufficient to avoid seeing any or even talking about them. If you are afraid of tight spaces (claustrophobia) or vast spaces (agoraphobia), to feel reassured you need only leave the door open in the bathroom in the first instance and avoid big shopping malls and the Grand Canyon in the second. You can also be afraid of an idea: the fear of throwing yourself out the window, for example (fear of an impulse); you will then be obliged to keep away from windows and to make sure that they are closed when you enter a room. The phobic neurosis, which bears certain traits of the anxiety neurosis and certain others of the hysterical neurosis, can also clearly be placed among the emotional disorders, and ascribed to the second brain.

4. The *hysterical neurosis* resulted in a great deal of ink being spilled and has fascinated psychologists and psychiatrists in all eras. For Henri Ey, "Hysteria is a neurosis characterized by the somatic hyperexpressivity of ideas, images, and unconscious affects. These symptoms are the psychomotor, sensorial, or vegetative manifestations of this 'somatic conversion.' That is why, since Freud, this neurosis is called conversion reaction."[2] This definition alone already poses the problem of a dialectic between the first and second brains, wherein the latter is in a certain way tasked with expressing and staging "ideas, images, and unconscious affects." Let us note also that the word "image" would be translated in my mimetic language by the word "model," and then we find the dialectic among the three brains: the first (ideas), the second (affects), and the third (models, obviously mimetic ones).

At this stage of our reflections I think there is a clear need for a new psychiatry that does not locate psychoses and neuroses in one of the three brain functions but accounts for them through the interaction of all three. This new way of seeing things will entail the reorganization of semiology and nosology and, of course, of psychotherapy.

PART 3.

An Essay in Mimetic Nosology

We are now going to attempt to reread classical nosology from the point of view of the third brain, that is to say by studying the state of the interdividual rapport for normal, neurotic, and psychotic experience, when the other is taken as a model, a rival, or an obstacle in the mimetic interaction. Here is a table representing this rereading, which we will study in detail, taking one square at a time.

	Model	Rival	Obstacle
Normal structure The other is real and recognized, but forgotten; forgetting and silence at N and N'	Recognition of difference; learning, identification, hypnosis	Vengeance Envy Jealousy	Renunciation or substitution of desire
Neurotic structure Real and misrecognized other; claim at N	Pathomimia Stigmata Possession Adorcism Mythomania	Diabolical possession and exorcism Hysteria	Resentment Psychasthenia Obsessions and compulsions
Psychotic structure Virtual other designated, accused, suspected, or hallucinated; claim at N'	Paraphrenia	Paranoia Chronic hallucinatry psychosis	Schizophrenia (fragmentation, dispersal)

CHAPTER 9

Figures of the Other in Normal Experience

The Other as a Model

Imitation and identification can be perfectly normal mechanisms when the model remains a model and the imitator lays claim neither to his belongings, nor to his being, nor to the precedence of his own desire with respect to the model's desire, nor to the ownership of that desire. The most famous imitators have no need to be jealous of Charles Aznavour or Michael Jackson: for the time it takes to sing one of their songs, they *become* Aznavour or Jackson; they imitate the model and identify with it. Impersonators of political figures add to this "peaceful" imitation a humorous dimension that does not reproduce the model but caricatures it by making it utter absurd or ridiculous things with its own voice, gestures, and distinctive way of speaking.

What makes it so that theatrical imitation remains "normal" is that it shows itself for what it is. By changing characters, voices, and gestures several times, the imitator allows us to see what we do all the time, without realizing it: it makes manifest the "wardrobe of personalities"[1] from which we constantly take our reactions, our attitudes, and our feelings. Professional imitators stage our otherness and our multiplicity.

In order to understand the importance of forgetting in the functioning and very existence of the self, we can take as an example the difference between an order and a suggestion. When you order someone to go into the street and open his umbrella, he takes you as a model and fulfills your desire without claiming that his desire has priority and precedence. As he descends the stairs and walks to the indicated intersection, he *remembers* at all times the order he has received, the person who gave it, and the action to be performed. In other words, the order does not create a new self in the person who obeys, whereas the process of suggestion-imitation does create a new self built on the overlooking of the otherness of desire. The order does not inject the desire in the other; it just imposes a will on somebody else. If the umbrella-opener is carrying out an order, this order may even be carried out against his own will, because he will think that this order is stupid and irrelevant because it is not raining. And he may even resent having been given this order, but he cannot avoid obeying it because of the social situation he is in.

By contrast, suggestion injects the desire into the other, who forgets where the suggestion comes from and is persuaded that his own desire that he is carrying out. If during the hypnotic trance, you give a posthypnotic suggestion to the same subject and suggest that when he finds himself in the street, at that particular intersection, he must open his umbrella, for that suggestion to be realized, he will, from the time that suggestion is made, have to *forget* or possibly misrecognize or negate the priority and the precedence of the other's desire to claim ownership of that desire without understanding where it comes from. And he will do all this so that when he performs the action and opens the umbrella, he will be the first to be surprised and will wonder why he did so.

◆ ◆ ◆

Forgetting preserves harmony in the self and keeps it in existence. We are all sleepwalkers from this point of view: we receive suggestions and we function in a sort of state of forgetting that helps us to live and that explains why we wonder so often what caused us to do such and such a thing. But this forgetting is relaxing, easy; it encourages laziness and is of course opposed to the difficult work that consists in recognizing the otherness of one's desire and the anteriority of the other's desire, which inspired it.

This forgetting is also manifested in collective psychology. In a crowd, mimetic attraction is hypertrophied and each person's self dissolves.

Everyone's desire melts into and is subsumed to the collective desire. Here psychological mass is the same as the number of people. Human beings gathered in a crowd are in a state of somnambulism, a plural somnambulism that has nothing to do with sleep. This somnambulism takes hold immediately and brutally, without going through the clinical stages of lethargy and catalepsy,[2] in my opinion because the mimetic hypertrophy is such that what Gustave Le Bon calls the "soul of the crowd" is immediately substituted for the dissolved self. This other self does not have to pass through the genetic stages of its constitution, because it already exists as a social, collective self.

Collective self of a collective desire, which is to say of a purely mimetic, contagious, irresistibly attractive, violent, labile desire; plural somnambulism, as I said, merging of desires, mimetic hypertrophying, dissolution of each person's self—such is the crowd. Plural somnambulism is "clinically" described by Gustave Le Bon: "Extinguishing of conscious personality, predominance of unconscious personality, *suggestion* and *contagion* of feelings and ideas in the same direction, tendency to immediately transform suggested ideas into action—such are the principal characteristics of an individual in a crowd. He is no longer himself, but an automaton that his will is powerless to guide."[3]

Rather than their will, it is in fact desire, mimetically vacuumed up, that can no longer claim any priority. Le Bon, prisoner of a psychology of consciousness and fascinated by Freud, characterizes plural somnambulism by the predominance of the unconscious personality. But it is obvious to me that what predominates is the other's desire, mimetic desire. The unconscious is the other, the Other, or the others. Consciousness, for its part, is a work in progress, a crossroads of information, a mimetic turntable, a temporarily present otherness, a temporary and unstable equilibrium.

There is lacking in what I have called "plural somnambulism" an essential aspect for it to be recognized as authentic somnambulism: amnesia. Nobody, to my knowledge, has clearly indicated that plural somnambulism is accompanied by a lacunar amnesia, one that is always partial, and sometimes total, but of course limited to the specific sequence of events generated by the crowd. Now the human being caught up in a crowd is "outside himself," as Philippe de Felice puts it;[4] he is "unconscious," Le Bon says. Let me repeat: we are dealing with a "psychological crowd" (Le Bon), a "delirious, delusional crowd" (de Felice) and not with the random, Brownian movement of the main hall at Grand Central station.

Once the psychological crowd is dispersed, once the "delirium" is calmed down, each self is reconstituted and amnesia manifests itself. It can be total: "What happened?" But more often it is partial: "How could I have done something like that?" How did I come to be covered in blood, with my clothes torn; how is it that I am covered in dust?"

To confirm this amnesia, we need only evoke the always divergent accounts of individuals who have been caught up in a crowd event. The objective falseness of their narrative proves the existence of their amnesia. Let us listen to Le Bon on this subject: "Collective observations are the most erroneous of all . . . innumerable facts prove the complete lack of faith one must have in testimony made by crowds. Thousands of men witnessed the famous cavalry charge at the Battle of Sedan, and yet it is impossible, in the presence of the most contradictory visual testimonies, to know by whom it was commanded."[5]

Amnesia that accompanies a plural somnambulism is thus, it must be affirmed, as real as that which accompanies a singular somnambulism, even if it is not as complete. This amnesia is, as in certain memory-related illnesses, compensated for by false reminiscences and confabulation.

Thus, sleep or the modification of the state of consciousness is not in the least necessary to the understanding and still less to the existence of hypnotic and somnambulistic phenomena, nor to their explanation. The cause of the latter, in the singular and in the plural, is solely mimetic.

◆   ◆   ◆

Getting back to individual psychology, the question arises: Why not remain in a state of relaxing oblivion? Because that state of forgetting assumes that the model remains a model at all times. Yet the least opposition on the model's part and the incapacity of the subject to handle this opposition will transform the model into a rival or an obstacle in the eyes of the imitator, as we will see, leading to a plunge into neurosis or psychosis. Forgetting is an Edenic form of the interdividual rapport before the fall described in the biblical Genesis, that fall being due, as I have shown elsewhere, to a rivalrous evolution of the mimetic rapport.

Forgetting presides over the constitution of the self at the moment of its creation. And it is certain that Adam and Eve had no memory of the moment when God created them by breathing onto the dust—just as we have no memory of the mimetic processes when we have been molded by the desire of our parents, who taught us to speak, to walk, to eat and made us into what we are.

Once there is rivalry, an opposition or an obstacle, a form of brutal awakening is manifested that the Genesis text presents as an expulsion from Eden and a projection into the world of difficulty and conflict. That rivalry or obstacle must then be explained. And to explain and try to come back to the Edenic state where our desire "belonged" to us, the otherness of my desire and the priority of the other's desire must be actively denied.

The Edenic, paradisiac state is characterized either by pure forgetting or by a peaceful recognition of the difference between the self and the model and of the precedence of its desire. The child does not know how to speak, and it is by modeling itself on its parents, by imitating them, by repeating better and better the words that they pronounce, that he learns the language. But when, for the first time, he manages to repeat "Daddy" or "Mommy," what joy! The parents are delighted and the child, congratulated and kissed, is too. The disciple thus copies the model, but the model wishes to be copied, and the same goes in all situations of apprenticeship. Later, the child will forget his babblings and will quite naturally speak "his" language. He will forget to which parent, or to which older brother or sister or teacher he owes such and such an expression, such and such a gesture or way of holding himself, such and such a habit. Here appropriation is normal and peaceful because it is consensual and desired by the model, and the forgetting of the origins of these acquisitions maintains the existence of the self in a state of serenity.

Besides imitation, which bears essentially on the model's appearance (his gestures, behaviors, reactions, way of speaking, and so forth), mimesis can bear on the model's very being. This is what all authors starting with Freud have called "identification," whose central role in psychology they have recognized. I think that identification in the sense that I have defined it above (see Part 2) is even more important than has been thought, for it is a physiological, normal, and "therapeutic" process aimed at avoiding the conflicts that result from appropriative mimesis: identifying with a model is consolation for not possessing all his or her belongings, since one *becomes* that model.

A striking example is apprenticeship in the manual trades. For here there is recognition by the apprentice of the superiority of the master who teaches him a particular art or technique. This, for example, is how knowledge was transmitted between the builders of cathedrals. The very fact of having one's desire identified with the master's in an open, knowing way and of seeking to surpass oneself or even to surpass him generated a production that modified the raw nature of the stones.

Echolalia

A curious phenomenon also deserves our attention here: in certain neurological illnesses, in particular in certain forms of dementia, the patient appears unable to respond to the doctor's question—he can only repeat it. The doctor says to the patient: "How are you?" and the patient repeats: "How are you?" This is what is called echolalia, which is often accompanied by echopraxia (repetition or imitation of gestures). The same phenomena were observed by Pierre Janet in hypnotic subjects in a state of catalepsy.[6]

This suggests that when the first brain is put out of commission by a neurological illness or hypnotic trance, the third brain is in a certain sense liberated from its influence, and the subject can no longer do anything but respond to a suggestion (verbal or gestural) with pure and simple imitation, without being able to add or subtract any information to what he or she observes.

By the same token, in certain other neurological illnesses, it is the second brain that seems to emancipate itself and produces laughter or tears that are said to be "spasmodic," which is to say without any affective or cognitive signification, and solely expressing mechanical activity of the second brain.

A Close Relationship among the Three Brains

There is thus a close and constant relationship among the three brains, which are intimately imbricated and interconnected by billions of neurons. It is these relationships that the new psychiatry must never lose sight of, trying in each case to determine where, in which of the three brains, the psychological disorder was born in order to target the therapeutic response and choose from among the medications and various psychotherapies those that will be the best able to help the cognitive, emotional, or mimetic malfunction.

This close relationship among the three brains is highlighted by the following anecdote. The great American psychiatrist Milton Erickson, inventor of a new form of hypnosis that bears his name and of new psychotherapeutic techniques, had an associate and friend who became mine as well, Ernest Rossi. I invited Rossi to come to the Sorbonne to do a demonstration for my students in the doctoral program in psychology. We asked for a volunteer and Françoise, a brilliant student who was doing her dissertation with me, raised her hand. The hypnotic induction was rather unorthodox: Ernest Rossi and I were seated on either side of Françoise. Rossi spoke a sentence in English and I translated it immediately as faithfully as possible, since Françoise didn't speak English and Rossi didn't speak French. Françoise entered the hypnotic state and revealed that she was tormented by a project that she was unable to finish because she was "stuck" on a point that was giving her difficulty. I understood that she meant her thesis, but I said nothing and translated faithfully what Rossi was saying to her: "You will manage to overcome the difficulty and finish because your unconscious will help you by finding the solution to your problem."

Françoise snapped brutally out of the hypnotic trance, in a state of great anxiety. Rossi and I were astounded. We tended to her and she calmed down. We then repeated what had been said during the trance because Françoise had of course forgotten, and it was then that she cried: "But you are crazy! My unconscious will never help me, all it can do is play tricks on me."

For several years François had been in psychoanalysis to aid in her training. For her, the unconscious was the Freudian unconscious, which her first brain had recorded as the reservoir of repressed desires and a source of problems. Rossi, for his part, was speaking of the Ericksonian unconscious, a benevolent and protective guardian angel, charged with helping the subject in difficult times. These two different conceptions had collided in Françoise: her first brain, violently shocked by Rossi's suggestion, had alerted the second brain, which had reacted with acute anxiety, and both had alerted the third brain, which had broken the hypnotic interdividual rapport.

In the same manner, no hypnotizer can make a subject under hypnosis perform an act that is contrary to his fundamental principles or beliefs. These astonishing phenomena show that the three brains are in constant interaction and notably that certain mental disorders or neurological or psychiatric syndromes are due to one of the three brains "breaking away" from the two others for organic or psychological reasons.

The Other as a Rival

When there is a rivalry but the model remains a model, the rivalry takes place in full awareness and recognition of what it is. It does not take refuge in the negation at points N and N', in other words in the claim to ownership of its desire and the priority of this desire over the desire of the other, which inspired it. Examples of this kind bear witness to a rivalry that I would call "constructive," because this rivalry really isn't one; it is an emulation in which the protagonists do not seek to destroy each other but to extend the limits of their mutual ambition.

For example, runners and swimmers seek to make faster and faster times in the hundred meters. They seek to beat their own record. And the athlete recognizes that he has a competitor, but at the same time that this competitor is also, in a sense, his or her coach: this competitor pushes him to surpass himself and does him a favor because he helps him improve his performance. The same goes for all emulations and all competitions in which competitive energy, instead of being directed at the "rival," is put in the service of performance.

In certain other sports, boxing for example, it might seem that this is not the case. True, the rivalry is ritualized, but the rival is the yardstick of one's performance and the performance is not good unless the rival is knocked out. And yet, even in a boxing match, once the combat is over, the second brain can provide feelings of admiration and friendship capable of transforming the rival into a model and the two boxers can then become the best friends in the world. That is what happens in the film *Rocky II*, in which the young Rocky admires the great Apollo Creed so much that he trains arduously to be able to face off against him. Apollo Creed, touched by this admiration, takes a liking to the young man, and agrees to fight with him, thus raising Rocky to his level. Rocky wins, but his admiration and his affectionate sympathy for Apollo survive their combat and are strengthened by the fact that the latter becomes Rocky's best friend and trainer. What is nice about this story is that, as opposed to the pessimism that emerges from Girard's oeuvre, it leads us to think that mimetic desire can bring about great actions, great feelings, and chivalrous situations instead of situations of resentment.

The same phenomenon occurs in the artistic domain. In his time, Plato saw that the artist, and notably the painter, is a copier of nature. I said earlier that imitations in cascade—which are in fact successive inspirations for generations of painters, musicians, and writers—are always an acceptance of the model, a recognition of the predecessor's desire as a model of one's own desire, but also a new production stemming from the inspiration provided by an ancient model. That is why there is a history of painting, of music, and of literature: if the phenomenon of mimetic imitation and change didn't exist, there would be no historical continuity.

It is clear, too, that the movements in the history of music engender each other mutually. Brahms, for example, recognized his model in exemplary fashion when, asked about the kettle drums of his First Symphony, he declared: "Those are the heavy steps of Beethoven striding through my music."

As a counterexample, we can evoke the wayward careers of certain Romantic poets who refused to recognize their model as such or that model's desire as the inspiration for their own. As a result, they made frenetic claims to originality. This claim to originality, which is found very often and in many cases, prepares the way for resentment. Even if it produces a literary work, it always leads the author or artist to misfortune or unhappiness or both.[7]

All of this leads me to specify here and now that a mimetic movement in the third brain gradually slips from the model toward the rival and then the obstacle. Desire, too, can gradually slip from the recognition of its inspiration to the claim to its precedence. As for the self, it is rare for it not to be persuaded of being the owner of its desire, but it can claim this ownership with greater or lesser virulence, driving it gradually toward neurosis or psychosis. Mimetic psychopathology will thus be dynamic, evolutionary, and cinematographic, while recognizing that by taking photographs of the process at various moments, those photographs may represent a neurotic or psychotic "structure" corresponding to a diagnosis. All of this must be seen as a traveling shot: everything is linked together, but if one selects film stills, neuroses or psychoses will appear as frozen individualities.

I think that in the framework of normal experience, when the third brain takes the model for a rival, it does so in full awareness of the superiority and the anteriority of the latter's desire, which is adopted and copied by the student's desire; then the model's achievements become a challenge. There is then, in a certain way, a form of identification with the model because one wants to do the same thing as him—as well as or better than he does it. The second brain then provides positive feelings: admiration for the model, and even affection, and the joy and euphoria of sharing the competition with him. This is particularly obvious in major tennis tournaments or matches when the two finalists, who are international champions, show a mutual admiration for each other. I recall the tennis champion Andre Agassi recounting how he lay, exhausted, in the training room after a particularly grueling but thrilling match, clasping his exhausted opponent's hand in recognition of their shared achievement.

As for the first brain, it has no need to provide rationalizations and simply gives its blessing to the whole undertaking while taking care of logistics (scheduling, preparation, organization, trips, and so on).

This is equally true when René Girard analyzed in a critical way the works of Freud and Lévi-Strauss. These two great geniuses were for him at once sources of inspiration and emulation, and if he offered different perspectives than them on ethnological and psychological reality, he did so with feelings of respect and even of gratitude with regard to these authors. Here we can cite the metaphor used by Bernard de Chartres, a Platonic philosopher of the twelfth century, and later taken up by Newton: "If I have seen further, it is by standing on the shoulders of giants."

But there is still more to say on this subject: desire itself recognizes the anteriority of the model's desire and claims to be closer to this desire than are the model's epigones. Thus, for example, in wishing to create a new metapsychology in tune with developments and discoveries in the neurosciences, developmental psychology, and so forth, I have the feeling of being completely in harmony with Freud's desire, which consisted in understanding the springs of human behavior—I seek nothing else. The psychoanalysts, for their part, copy Freud's conclusions, Freud's productions, Freud's analyses, but never does their desire merge with Freud's, which was to discover the reality of psychological mechanisms. They imitate his teaching and copy his conclusions, without truly adopting his desire for discovery.

In fact, when the model is a rival in normal psychology, everything is played out in a subtle balance between the third and second brains: if, for example, an athlete admires another who performs better than he does and he *loves* him, well then he will take him for a model and improve his own performance; but if by misfortune he develops feelings of envy and jealousy for his competitor, then he will be gnawed at by mimetic rivalry and will never improve his performance, deprived as he will be of the training/coaching effect, and thus of the feelings of admiration and friendship that the fact of taking the other as a model could have on him. Thus, envy and jealousy, even if they occur in full awareness of my desire's otherness, have a weakening effect that slows my progress. Curiously, envying one's rival makes it impossible to imitate him correctly.

Finally, rivalry can be experienced in full recognition as deferred violence, that is to say as vengeance. Greek tragedies—and closer to us a tragedy like *Le Cid*—recount nothing else: although the models on whom the hero wants to take vengeance are fully recognized as admirable and respectable, revenge requires fighting against them, treating them as rivals, and killing them, which dismays the hero—and his second brain complains:

> The poor avenger of a cause that's just,
> Compelled by unjust fate to play that part,
> I stand here stunned, and with a head held low
> Yield to the fatal blow.[8]

There are rivals against whom one can fight, but there are others against whom nothing can be done and who constitute a sort of transition toward the model-obstacle, which we will speak of later on. An example of this is James Joyce's short story "The Dead." In this story Gabriel and Gretta are married and, going back to their home, share some tender moments. Gabriel's desire, which is perceived and felt by Gretta, is communicated to her and colors her mood with tenderness and acceptance. Gabriel is thus sure of his "victory," to the point that he wonders why he waited so long to take advantage of his wife's willingness. But suddenly he observes that Gretta is pensive and he asks her questions to find out what she is thinking about—let us remark in passing that the most unexpected events can occur when we take the risk of asking someone on the spot what he or she is thinking. Gretta declares that she is thinking about a song that they have just heard and at once she detaches herself from Gabriel, runs to her room, and throws herself on the bed, hiding her face in the cushions.

Dumbfounded, Gabriel follows her and asks her why she is crying. "I am thinking about a person long ago who used to sing that song," says Gretta.[9] Questioned about his identity, she replies that it is a person that she knew long ago in Galway. This unexpected and potential rival makes Gabriel angry. He tries to ask Gretta ironically if she was in love with this person. "It was a young boy I used to know . . . named Michael Furey. He used to sing that song, *The Lass of Aughrim*. He was very delicate."[10] Gabriel no longer knows quite what to say and does not want to reveal the interest that he is taking in this "delicate" boy. Gretta puts the finishing touches on his discomfiture: "I can see him so plainly. . . . Such eyes as he had: big, dark eyes! And such an expression in them—an expression!"[11] The conversation continues and Gabriel learns that Gretta took walks with this young man when she was in Galway—Galway, exactly where she has asked her husband to go on a trip with her: was that to see the boy again? And now Gabriel is stunned by Gretta's reply: Michael Furey is dead. "He died when he was only seventeen."[12]

Gabriel feels humiliated by the failure of the irony he attempted to employ, but the idea that his wife could compare him to another in her secret thoughts makes him sick with shame. And yet he does not give up and wants to know everything: "And what did he die of so young, Gretta? Consumption, was it?"[13] And here Gretta delivers the coup de grâce: "I think he died for me,"[14] she says.

Gabriel is overcome by terror: he is beaten, totally helpless when faced with this rival who is dead and gone and who seems moreover to have given his life for Gretta, something that he will never be able to compete with himself. This dead rival has thus triumphed without even putting up a fight, and one fears that henceforth the young Michael Furey will constitute an obstacle for Gabriel, an insurmountable obstacle between him and Gretta. The infinite or absolute outcome of desire is tragic: to recognize that one can never fill in the gap between oneself and the model-rival is to realize that the model can never be reached; thus does it become an obstacle.

The Other as an Obstacle

Faced with the obstacle, which is to say the rival who is always already there, who is so to speak insurmountable, the "normal" attitude consists in renouncing competition with the other and in redirecting one's desire in a more constructive direction, in accepting one's own limits as well as those imposed by the social structure (for example, by the law). Renunciation is at the foundation of all hierarchical societies. This is because renunciation shows desire its impossibility. The obstacle is there, absolute. The mediator is always external, even if he is right in front of you.

For example, Colbert was a very powerful minister, but it never occurred to him that he could become a duke and compete with a great noble of the kingdom, for ancien régime society was divided into the nobility, the clergy, and the third estate. And the members of the third estate could not imagine, even in their wildest dreams, appropriating the titles of the nobles. When, in the ancien régime, a man, no matter how powerful and intelligent he might be, took it into his head to lose sight of the fact that the king was an insurmountable obstacle and put himself in the situation of appearing as a rival in the king's eyes, he ended up in the Bastille. This was the case with Nicolas Fouquet, whose motto was *Quo non ascendet* ("What heights will he not scale?"). Louis XIV took it upon himself to clearly mark the upper limits of Fouquet's ascension. We can also mention Jacques Coeur, whose fortune aroused the jealousy of Charles VII, as Jean-Christophe Rufin shows in an admirable novel.[15]

On the individual level, the child himself can in no instance structure his personality if he is not shown that the rules admit of no debate, that they are nonnegotiable, impossible to get around, and thus insurmountable. Today's parents, if they give up embodying those rules, fall into rivalry with their children, a rivalry that is disastrous for their development and very painful for the parents on a daily basis. For these are parents who, being in a position of rivalry because they put themselves on the same level as their children, are obliged to justify all of their requests and all their decisions and, worse still, to negotiate their application. This is what I observe every day.

It is as if, in the Garden of Eden, when they wanted to eat the fruit, God had been summoned by his "children," Adam and Eve, who demand that he explain himself and his decision to forbid the tree in question rather than another one. This shows that rivalry is always the product of a destructuration of society leading to a weakening or disappearance of authority. Starting from the moment when authority is obliged to justify itself, the father puts himself on the level of the son and allows him to enter into rivalry. In the myth of Oedipus, it is the fact that Laius, in his cart, puts himself on the level of Oedipus by giving him a blow with his whip so that he will clear the way that allows the son to kill his father. If, to the contrary, Laius had draped himself in his dignity: "Young man, I am the king, I am old enough to be your father, move aside," Oedipus would never have thrown himself on him. Laius lost his prestige by wanting to replace authority by violence.

This leads us to note in passing that violence cannot be exercised against an obstacle and thus cannot be exercised against an external mediator who lays down a rule. Adam and Eve can disobey, but they will be punished and would never have the idea of attacking God. When the rule is broken, the second brain generates feelings of fear, regret, repentance, and humiliation. In a word, of guilt. Guilt is thus the second brain's reaction to a mimetic behavior of disobedience with respect to prohibition, that is to say to a mimetic behavior that has lost sight of the obstacle, and thought it could get around it.

Some prohibitions are not taken seriously and their transgression implies no feelings of guilt on the transgressor's part. I am thinking of the patient who told me that she had run a red light at one in the morning while going home. To her great surprise, a policeman appeared and stopped her on the other side of the intersection. The policeman said to her, "So, Madam, you didn't see the red light?" And she replied: "Yes, but I didn't see you."

In this case, transgressing the prohibition entails no guilt feelings, but to the contrary a euphoric reaction of the second brain, which is only transformed into remorse when the transgression is discovered and punishment looms.

Other prohibitions are to the contrary so profoundly rooted in us that they bring forth remorse even if the transgression is never discovered. Freud had a word for this external mediator, which proclaims the prohibitions and which, being located in the very brain of the one who disobeys, makes him unhappy and full of remorse without external interference: the "superego." This is the case of Raskolnikov in *Crime and Punishment*, whose superego visibly obliges him, in order to find peace and serenity, to confess his sins and transgressions even though he has not been caught red-handed (the expression is literal in this case). In Dostoevsky, it is clear that the superego is one with what the Christian tradition calls the "conscience," that is to say the reference to God's commandments, to the prohibitions proclaimed by him, and to final judgment. This guilt that permeates Dostoevsky's oeuvre runs through the whole history of Judeo-Christianity. On the other hand, in his novel *The Kindly Ones*, Jonathan Littell presents us with an abominable scoundrel, a killer who, having no superego, no faith, and thus no Christian "conscience," serenely pursues his life by taking advantage of all the benefits derived from his crimes. Throughout the book, Littell's hero experiences no anguish or any form of guilt except when he feels that he has been found out and is sought by the police.

The Freudian father represented throughout the duration of life by the superego was a categorical and insurmountable obstacle. Now this kind of father has disappeared, and with him the superego, it seems, such that our society is obliged to multiply the number of laws and rules in an attempt to predict all possible infractions in all areas. In the ambient cacophony, justice itself continuously becomes subject to negotiations, challenges, and legal nitpicking, it being understood by the lawyers on both sides that every argument can be countered by another argument. Justice is dissolved into litigiousness when it proves incapable of replacing the father's categorical imperative and unquestioned authority.

The prototype for a declaration that denies the model as obstacle is Napoleon's remark: "Impossible is not French." This phrase left its mark on the collective unconscious of my compatriots, who have gone so far as to no longer see the obstacles around them and to declare that it is "forbidden to forbid."

CHAPTER 10

Figures of the Other
in Neurotic Experience

The Other as a Model

Under this heading, the first case is *mythomania*, or pseudologia fantastica. Mythomaniacs are people who tell stories that are nothing but lies, and they really are lies, because unlike the paraphrenic, the mythomaniac does not believe them. He may, for example, present himself as a great businessman, and he tells a whole story that has nothing whatsoever to do with reality. He thus identifies with a model that makes him look good and he momentarily appropriates the desire of that model, but in the form of something borrowed. If he borrows this desire, this is for the needs of the narrative with respect to a third party and not at all because he is appropriating it for himself in a delusional way. Mythomania is a form of self-aggrandizement and deception, in the final analysis.

Things become more serious in the case of *pathomimia*. Here it is a matter of making the story that one is telling more credible by presenting cutaneous lesions that one has made oneself for the sole means of pretending to be a sick person, that is to say a victim. It is a way of attracting attention, the sympathy of other people, and of making oneself interesting, the lesion being meant simply to lend credence to the reality of the story that one is telling. This is the reason why these cases of pathomimia are most often due to self-mutilation. One must be wary of extremely frequent slippages into something more serious, for self-mutilations can indeed be the expression of a pathomimia but very often, too, of a psychosis.

Patients can also use doctors and above all surgeons to give credibility to their illness. This is the case notably of a patient who presents a Munchhausen syndrome who arranges to be operated on by mimicking the symptoms of an intestinal occlusion, for example, knowing that the surgeons will react by putting him or her under the knife.

Finally, we cannot fail to mention the case of *stigmata*. Those who bear such wounds are clearly imitating the wounds of Christ. In this way they portray in their flesh the passion of Jesus and as a result lay claim to what they believe is his desire. This extremely rare phenomenon bears witness all the same to the power of imitation, as well as to the absolute unity of body and mind, contrary to what Descartes may have thought. The narration and reminiscence of the Passion of Christ appears throughout the centuries in the canvases of painters, but also in the living paintings constituted by those with stigmata. It is said that stigmata occurred only after the thirteenth century when Western painting first depicted the sufferings of Christ and the crucifixion. In that sense those stigmata could be considered a secondhand imitation—the imitation of the painting rather than the imitation of Christ himself, whom they (those with stigmata) have never seen.

This portrayal is self-serving but obviously theatrical. And that theatricality is typically neurotic. This suggests that in neurotic phenomena, the "clothing" is essentially taken from the second brain's wardrobe, which is to say from the emotions and feelings.

From this point of view, the psychoanalysts have always been in the best seats in the house of this neurotic theater, and they have given utterly fascinating reports by endlessly analyzing these performances and seeking to get back to a historical meaning and a causality in the very history of the subject. I think to the contrary that whatever the "getup" and the staging, it is always a matter of the self's claim at nodal point N:

· In *mythomania*, the "actor" integrates—for the duration of his narrative—the desire of his character in order to make himself look good and impress his interlocutor. The neurotic enterprise is there, present for as long as the discourse lasts.

· In *pathomimia*, the desire to be sick and to have an illness becomes frenetic and bloody. The subject is ready to experience all kinds of pain and suffering to lend credence to the fact that he is sick. What is paradoxical is that this desire to be sick is authentic but denied as such, since being healed is demanded of the doctor, and the medical system is thwarted by the very fact that the patient is using all means to stay sick. What makes this behavior a neurotic behavior is also the fact that the patient, deep down, knows very well that he is not sick and that the sickness is just the image of himself he wants to give others. The skill of the doctor will consist in making him become conscious that although he is not sick with the sickness he is pretending to have, his real sickness consists in wanting to be sick. Furthermore, these patients, like many others that we will encounter in practice, will take the doctor as a rival and make sure to stay sick in order to prove that he has failed at the challenge of healing them. This goes to show that there are permanent overlap and transitional stages between the model as model and the model as rival.

· As for those with *stigmata*, they appropriate the desire of their model so violently, they so want to suffer like him that the images of that suffering end up being inscribed on their skin. They really do bleed. But what makes their case fall under the neurotic heading is that they never mistake themselves for Christ and are quite conscious that their wounds are only copies of their model's. Let us note besides that the truly delusional people who take themselves for Christ never have stigmata. Without wanting to venture into theological territory, this perhaps shows that Christ never desired his Passion but only accepted it. Finally, in the little theological excursion we have just made, we may also be struck by the fact that the punishment of the cross lends itself admirably to all representations. The cross is a symbol, the open arms too, and also the wounds. Jacques Corraze recalls: "The wounds of Jesus, says Saint Bonaventure, are the flowers of blood of our sweet and flowery paradise where the soul like a butterfly floats, drinking now from this one and now from that."[1] All that the cross represents for billions of believers would have been impossible if Christ had been put to death in a different manner, even if his head had simply been cut off like John the Baptist's. Is there a divine design in the choice of the place of the incarnation and thus of the punishment of the cross? That is one more mystery that we must ponder. The stations of the cross, the exposition, the resurrection, and so forth—all of this is such a successfully realized theatrical spectacle that one is tempted to applaud the director.

The Other as a Rival

Faced with another who is perceived as a rival, the first reaction is fear. If the rival is neither perceived, nor singled out, nor recognized, the fear will seem to be without an object. This is what happens in the anxiety neurosis. But this situation of "fear of nothing" or "fear of fear" cannot be accepted by the first brain: the latter needs to rationalize the perception of the other as a rival and to justify the explosions of paroxysmal anguish against a backdrop of diffuse anxiety produced by the second brain.

It will thus be necessary to identify the rival. The latter is, by definition, misrecognized by the third brain, for at N' the ownership of "my" desire with respect to his or hers is asserted, and it is precisely this misrecognition and this claim to ownership that makes him or her my rival. Truly identifying the rival would be the objective of interdividual psychotherapy. But, for the moment, the neurotic enterprise is going to represent that rival.

In the first stage, this rival will be represented by an object, an animal, or a situation. "I am not anxious but I am afraid of mice and rats. It's physical. They disgust me." Or else: "I am not anxious but I am afraid of sickness, and above all of suffering." And the dedicated reading of medical books will then nourish the fear and rationalize it: "I have a headache; maybe I have a brain tumor," "I'm coughing, couldn't this be the first sign of lung cancer?" and so on. Or it can be "I am not anxious, but I am apprehensive about going out alone in a crowd or in a big store. I prefer to be accompanied." Or "I am not anxious, but I am afraid of knives, closed spaces, elevators, for example, and of high places because I get dizzy. I am even afraid of walking under a ladder.

I must be superstitious." This first way of representing the threat, that is to say otherness in its rivalrous dimension, instead of representing the rival himself, is progress in the defense mechanism against anxiety: one only has to avoid the object or the situation that triggers the phobia to avoid anxiety.

The next stage will consist in representing the other by "otherizing" a part of the body or a physiological function (sight, for example). This enterprise represents the other, otherness: it hides the other behind conversion symptoms,[2] but represents it during attacks of hysteria.

I have studied this neurosis at length in chapter 4 of *The Puppet of Desire*. What emerges is that this neurosis has its origin in the third brain, the interdividual rapport oscillating between the model taken as such and the model as rival, between love and hate. The fluctuations of this interdividual rapport are then staged and colored with various feelings and emotions by the second brain.

Before arriving at neurosis, that is to say at the organicization and the staging of rivalry while avoiding and dissimulating the rival, an intermediary situation can occur:

"I'm leaving you," says the young woman. "I feel suffocated by you. I need freedom, independence, and to take control."

"I am crushed," replies the lover. "I love you. I loved you so much. Nobody else, ever, will love you like I do. Is there another man in your life?"

"Yes. He is young, dynamic, he likes to go out, to dance, to laugh. With him I feel alive, I feel young."

The lover is silent. The young woman takes his hand gently, brings it to her lips, covers it in kisses, and says to him: "Forgive me. You are and will always be the only man in my life, the only one I truly loved. I will never kiss another man's hand."

Astounded, moved, overwhelmed, disoriented, the lover nonetheless gets ahold of himself, mobilizes his cognitive brain to give meaning to what is happening to him, and finally cries out: "I understand! There are two women in you. One is marvelous, divine, intelligent, reasonable, and she loves me. The other is a wild horse, impulsive, going with the flow, eager to gallop through the fields without a specific destination, intoxicated by her speed, her power, elated by her freedom."

This rationalization reassures the lover. Yet it corresponds to a different reality. There are not two women in her, but between them there are two types of interdividual rapport: a rapport where the model remains such, the model she loves and respects, grateful for all he brings to her life; and a rapport where the model becomes a rival, rejected, for although he is reasonable and loving, he is heavy, weighing her down, and the horse rears up and wants to break free.

When this duality of the interdividual rapport organizes itself, we enter into neurosis: the active and demanding misrecognition at points N and N' can only become organized, in the physical absence of the model-rival, by representing that model-rival. And the hysterical way of arriving at this result, as I have already said above, is to "otherize" a part of oneself to represent the other, a limb or a physical function in the case of hysteria, and we will see that in certain psychoses the model-rival will be represented by an "otherization" of a part of the psyche.

◆   ◆   ◆

Here we are in a situation where, at the level of the third brain, the interdividual rapport becomes conflictual and the model a rival. Let me here remind the reader of things I have already explained in *The Puppet of Desire* and (more briefly) in *Psychopolitics* about the various masks of the other and the various representations of otherness throughout the history of psychology.

The first attempt to represent otherness goes back notably to Hippocrates, to Plato, to Aretaeus of Cappadocia, who said basically: "There is in both men and women an organ that behaves like a physical Other in that it is endowed with a life of its own and launches initiatives that the self or the 'subject' can do nothing to oppose." In men, it is clear that the sexual organs do not obey the will, that it is the sexual organs that draw along their bearer and not the other way around. In the same manner, in women, the uterus is thought of as an animal with wings constituted by the fallopian tubes and the ovaries, and this animal, when it is irritated, rises to the upper regions of the body, which explains the hyperventilation, fainting fits, and wild movements of attacks of hysteria. Thus, this whole first part of the history of neurotic pathology, and notably of hysterics, is marked by an otherization of a part of the physical organism that behaves in an independent manner and draws the subject along in its wake, whether the subject likes it or not. This way of seeing makes it possible to lay claim to the ownership of one's desire, given that it is born within the organism, while avoiding assuming responsibility, since it is in reality triggered by an *intraphysical other*.

With Saint Augustine, the notion of an intraphysical other becomes sacrilegious inasmuch as every human being is created by God. An organ thus cannot be at the root of problems and escape the subject's control. On the other hand, the idea that gains popularity is of an *extrapsychic other*, the Devil, in various forms (incubuses, succubus, and so on, and later witches), who penetrates the subject's consciousness without his knowledge and leads him into temptation. This way of seeing things would not last very long, for in a sense it deprives the self of the ownership of its desire, since the latter was injected into the self from the outside by the demon. That is why, in the seventeenth century, Mother Jeanne des Anges, after having carefully thought over the events that took place during the so-called possession of Loudun, in the Ursuline convent, wrote that the Devil could have no influence over her unless she gave her consent and concluded: "I am manipulated, but I am manipulated by myself," thus making good at point *N* the claim to ownership of her desire.

Mother Jeanne des Anges opens the way to the third variety of interpretation that culminates in Freud, when he lends support to the claim at *N* by imagining an *intrapsychic other*, enabling the self to claim ownership of its desire but to avoid taking responsibility because it is unconscious. The responsible party is an intrapsychic other beyond the will's reach, inaccessible to consciousness, and fulfilling all the conditions for neurosis because the self is owner of its desire without being responsible for it and still less guilty of anything.

As I have already suggested, the hysteric represents the otherness of rivalry that is poisoning his or her life by otherizing a part of his or her body that from then on represents the rival inasmuch as it becomes the source of suffering: paralysis, parasthesia, burning sensations, pins and needles, hysterical blindness, astasia-abasia,[3] and so on.

A few examples from everyday speech: "He's a real pain in the neck/ass," "He's getting on my nerves," "She really makes me sick," "He's getting under my skin," "She's gotten inside his head," "You're smothering me," "You're driving me crazy," and so forth. All of these expressions translate a reality that the patient is expressing without completely realizing what he is saying. And, very often, it is enough for the psychotherapist to repeat slowly: "He's a real pain in the neck," for the pain that the patient is complaining of to receive a different interpretation.

Two examples come to mind. During the presidential elections in a little, French-speaking, third-world country, two candidates were running against each other in the final ballot. The National Assembly was divided more or less into two camps, but one of the candidates was a deputy and could thus vote for himself, which was not the case for the other. The first was elected by just one vote (in French *voix*, meaning both "vote" and "voice"), and it was clear that it was his own! I was called to examine the unfortunate candidate who had lost his voice (*voix*—voice/vote) since the defeat and who was completely unable to make a sound. Obviously, we first had to go through the ENT checkup and the fanciful hypotheses of various doctors. I ended up having a private conversation with the patient, during which we both realized that in fact the "voice/vote" that was lacking for him to be elected was his own and that his somatic manifestation merely represented reality, since he had lost because he was lacking his own voice/vote. In a way, his rival had taken it away from him. My patient was translating into his body the reality of what had happened.

The second example is that of a patient who was about fifty years old when he lost his mother, to whom he was profoundly attached. The very next day, his hands started to swell up and then to blister, with pustules and an eczema that made him suffer cruelly and that baffled the dermatologists. It should be noted that this man had always been a spoiled and spendthrift child, and that his mother had repeated to him many times that money had a habit of "slipping through his fingers" and "burning his hands" (meaning in French that he was eager to get rid of it, a bit like money "burning a hole in one's pocket" in English). The unfortunate fellow walked around wearing black gloves, and his social life was seriously disrupted. Not knowing what to do, I encouraged him to take a break and go on a trip in order to get over his grief. He chose to go to Israel, and while visiting the Dead Sea, someone suggested to him that he go in the water and especially that he soak his hands, which were covered with lesions and lacerations. He did so. The next day his hands had become normal, as if they had been regenerated by being plunged into the Dead Sea (in French *mer* means "sea" but also sounds exactly like *mère*, meaning "mother). Here again, reality was translated into the body in a very violent way. The rival of this patient was death, which had taken his mother. Plunging his hands into the Dead Sea / Dead Mother was for him a kind of resurrection, a regeneration, and perhaps also a form of substitute forgiveness.

◆ ◆ ◆

Now I would like to take an example from Robert Musil, who has described for us in his novel *The Man without Qualities* a scene during which Ulrich, the novel's protagonist, and Gerda, a young woman he knows, are going to bed to make love. Musil makes us eyewitnesses to what happens when the young woman is hysterical and frigid.[4] First, Musil underlines that Gerda thinks about making love with Ulrich with as much apprehension as if she were going to the dentist's office to have a tooth pulled. Ulrich perceives this: "He guessed that Gerda had made up her mind to get it over with because it was too late to get out of it." The interdividual rapport is immediately influenced by the reluctance of the young woman, which contaminates Ulrich: "His hands encountered the girl's skin, still bristling with fear, and he felt frightened, too, instead of attracted. . . . It made no sense to do what he was doing, and he would have liked nothing better than to escape this bed."

Musil then makes a digression to explain the cultural reasons why Ulrich persists in his undertaking. He shows him to us summoning up "all the usual reasons people find nowadays to justify their acting without sincerity, or faith, or scruple, or satisfaction"—in other words, all the reasons that men and women come up with today for engaging in utterly stupid behavior.

With such negative feelings on both sides, the interdividual rapport is so ill-used and abused that a fantastical rivalry sets in. Gerda expresses this and does not recognize the usual, nice Ulrich in "this naked stranger whose hostility she sensed and who did not take her sacrifice seriously." Faced with this enemy, this performer of sacrifices, "Gerda suddenly heard herself screaming," and the hysterical attack is triggered: her "lips grimaced and twisted and were wet as if with deadly lust. . . . Her eyes would not obey her and kept sending out signals without permission. Gerda was pleading to be let off. . . . Her hands were up over her breasts, and she was menacing Ulrich with her nails." What is noteworthy about this hysterical attack is that the rival does not have to be represented because he is present; nonetheless, Musil intuits the theatrical and histrionic aspect of the crisis, which mimes the rivalrous rapport: "She perceived it with utmost clarity as a kind of theater, but she . . . could do nothing to prevent her fate from being acted out before her, in a screaming frenzy; nothing to keep herself from taking the lead in the performance."

This crisis destroys all the positive feelings that might have existed in Ulrich, and Lacan would have had more reason than ever to say, *Il n'y a pas de rapport sexuel*, "There is no sexual rapport" (in other words, there are only human relations, which color and influence sexual ones). "At long last her fit began to wear off. . . . Ulrich took a deep breath, again overcome with repugnance at the inhuman, merely physical aspects of the experience."

What is remarkable in Robert Musil's narrative is that he shows the model becoming a rival beneath our very eyes and beneath the partner's eyes as well. Musil shows us, once more, the reasons for frigidity in hysterical women. The last lesson that Musil gives us is the transformation of this theatrical scene of rivalry into a medical rationalization: "Ulrich's voice sounded more tender; all it meant was that he regarded her as a sick person." And, of course, it does not escape Musil that the hysteric refuses and denies responsibility for his or her symptoms and illness, and accuses the rival of causing them: "But it was he who had made her sick!"

Hysteria is the neurotic expression of rivalry.

The Other As an Obstacle

When the third brain perceives the model as an obstacle, there are, it seems, two ways of reacting. The first consists in reacting with the equivalent of renunciation in a normal "structure," which French nosology, and notably Pierre Janet, has called *psychasthénie*. This illness—which today is no longer recognized for reasons that I cannot explain—was characterized by fatigue and slowing of all physical and psychological activity, as if, to make the least gesture or to take the least initiative, the subject had to fight against or push back against a force that stood in his way and prevented him from moving.

The second solution, when one encounters an obstacle—a closed door, for example—is to ceaselessly come back to rub up against the obstacle, knock on the door, ring the doorbell, scratch the door, and so forth. This is what happens in the obsessive-compulsive neurosis, where the subject is incapable of performing an action without repeating it an infinite number of times, for the obstacle prevents him from reveling in and acknowledging that the action is finally completed. Thus he washes his hands ten, twenty, thirty times in a row without this ever being enough, because his desire for cleanliness continually runs up against the obstacle constituted by the other, a bit like the indelible bloodstains on Lady Macbeth's hands, which guilt makes visible to her.

By the same token, in another variety of obsessive-compulsive disorders, the repetition of rituals, of counting for example, is compulsive in the same manner as knocking repeatedly at a door that doesn't open. Some other patients can only advance very slowly because they are obliged to take three steps ahead and two back.

The third variety of obsessive-compulsive neurosis is obsessive hoarding: the patients can't throw anything away and collect the most insignificant objects, to the point that this monstrous accumulation ends up filling their residence and expelling them from it, like the patient that we found dead on the doorstep in front of his house: the latter was so stuffed with a mass of newspapers going back fifty years that the beneficiaries of his estate decided it was easier to tear everything down.

In these neuroses, it seems that the third brain, continually coming up against the obstacle, gives to the first the task of rationalizing this perception of the other. In her play *The Unexpected Man*, Yasmina Reza's female character describes her brother who is convinced that walking in his hallway only on the white and beige tiles and avoiding the black ones is vital for the survival and equilibrium of the world. The second brain supervises the carrying out of rituals entailing disaster if they are not respected. And again, Yasmina Reza tells us that when the concierge crosses the hallway, stepping on the black tiles, an incredible anxiety submerges the above-mentioned brother, and he must close his eyes to avoid seeing such a monstrosity.

It is very difficult if not impossible to treat or heal an obsessive-compulsive disorder, because the other who constitutes the obstacle is almost never identifiable or is so integrated or goes back so far (for example, a relationship with the father or mother) that one cannot treat the patient except by treating the symptom itself, as cognitive therapy does.

The Waterloo Syndrome

I have often encountered patients who were obsessed with their own failure and who wanted to keep replaying the lost battle. I call that the "Waterloo syndrome," because the only general who ever lost that battle was Napoleon; all the others who have replayed it since, starting with Clausewitz, won it because they knew exactly what to do. Thus did one of my patients develop an obsessive-compulsive disorder after having left the company where she was working, because a colleague had impregnated her and then abandoned her. Twenty years later, she was still obsessed by this idea: "If only I hadn't quit my job and if only I'd been able to hold on to Hubert."

Her first brain reconstituted the story to show that she would have been perfectly happy if those two events had not taken place, and it was impossible to make her acknowledge that thinking only about that was precisely what was causing her illness and that her illness wasn't due to her resignation but to the fact that the latter had been enshrined as a definitive obstacle to all future undertakings or successes.

When this patient telephoned me, she would say: "The Waterloo syndrome has come back with a vengeance." In her particular case, the obstacle was not a single person; it was circumstances, the head of the company who should not have accepted her resignation, her boyfriend, and so on. This mutation of reality into an obstacle had been elaborated by the third brain, which had charged the first to, so to speak, put a hat on it, to explain it and make it last, while asking the second brain to dress it up with feelings of regret, remorse, and anguish.

I cannot close this chapter on the neurotic obstacle without citing for the reader this text by Kafka, from his novel *The Trial*, which so perfectly expresses the situation under discussion here. In this case, it is the Law that represents the obstacle that the man's desire stubbornly wishes to surmount, or at least to understand. But the obstacle is itself protected by a doorkeeper, who forbids access to it: "Before the Law stands a doorkeeper. To this doorkeeper there comes a man from the country who begs for admittance to the Law. But the doorkeeper says that he cannot admit the man at the moment. The man, on reflection, asks if he will be allowed, then, to enter later. 'It is possible,' answers the doorkeeper, 'but not at this moment.' Since the door leading into the Law stands open as usual and the doorkeeper steps to one side, the man bends down to peer through the entrance. When the door-keeper sees that, he laughs and says: 'If you are so strongly tempted, try to get in without my permission. But note that I am powerful. And I am only the lowest doorkeeper. From hall to hall, keepers stand at every door, one more powerful than the other. And the sight of the third man is already more than even I can stand.'"[5]

For years and years the man observes the doorkeeper almost without interruption. He forgets about the other doorkeepers. The first seems to him to be the sole obstacle. Finally, as he is about to die, the man asks: "Everyone strives to attain the Law . . . how does it come about, then, that in all these years no one has come seeking admittance but me?"[6] The doorkeeper, sensing that the man's end is drawing near, shouts into his ear: "No one but you could gain admittance through this door, since this door was intended for you. I am now going to shut it."[7]

Thus this obstacle was for him alone. For him alone indeed this obstacle was the absolute model, because it forbid access to something unknown but which, given how well it was guarded, must have been the most desirable thing in the world. It is the "man from the country," here in the role of the patient, who takes the obstacle for a model: here is the most formidable prohibition, here too must be the most absolutely desirable object. And the man spends his life rubbing up against this obstacle, trying to get around it, to seduce or buy it. And the doorkeeper accepts his gifts, while specifying that if he does so, it is so that the man will feel that he has done and tried everything, just as all OCD patients do.

Thus does the obsessive compulsive exhaust his life rubbing up against the obstacles that he chooses for models and, as Kafka puts it so well, which are model-obstacles only for him!

CHAPTER 11

Figures of the Other in Psychotic Experience

The Other as a Model

Here the model is an absolute model and what is psychotic in the enterprise is that the imitator does not identify with the model but claims the precedence of his desire with respect to the model's; as a result, he proclaims that the true model is him and that the other—a famous personage—is in fact nothing but a failed copy of himself. Physical and historical time is then, beneath our eyes, inverted by psychological time.

One day, when I was an intern at Sainte-Anne Hospital in May 1968, I was brought a Spanish engineer who declared to me that he was the real Napoleon. I told him that this surprised me because it was commonly believed that Napoleon had died in 1821. And then he claimed insistently and even with irritation that the real Napoleon was him, the absolute proof being that the "other" had been defeated at Waterloo, which would never have happened to him! Moreover, he said that the "other's" enormous mistake was to have invaded Spain, seeking to Europeanize it, whereas *he*, the true Napoleon, planned to Hispanicize Europe, which obviously seemed to him a much better idea.

This Spanish engineer was a seasoned connoisseur of the entire history of the Napoleonic Wars and I had the good fortune, being an avid amateur historian myself, to be able to discuss certain finer points that gave him confidence in me as an interlocutor. But this also enabled him to underline the errors committed by the emperor and to give me to understand that he would never have committed them, the proof being that he was criticizing them and could explain where the other Napoleon had gone wrong.

The instructions of my superiors concerning this patient were to gradually increase the doses of neuroleptics to see if we could obtain the surrender of this delusion with a single theme and a single mechanism, imagination. It was thus a paraphrenic delusion. Every morning, during my visit with him, I played the game. And the events that were unfolding in that month of May 1968 (the month during which the famous student uprising took place), afforded topics of conversation.

"Sire, what do you think of the situation in the streets of Paris?"

"I am not going to put up with it for very long, but for the moment you are keeping me prisoner, and you are preventing the French from rallying around their emperor. You are thus responsible for this unbearable insurrectional situation."

After three or four weeks, by which time the dose of neuroleptics was very high, I went into his room one morning and as usual asked him: "Sire, are you satisfied to see that the situation is calming down?"

He looked at me as if I were a half-wit and replied in a very pronounced Spanish accent: "You know, doctor, I'm Spanish. So I really couldn't care less about French affairs."

He was "healed."

This case is the only one I encountered in my career in which the patient took himself for Napoleon. And even then he had to be Spanish, because the French patients I treated were more likely to take themselves for famous singers or television stars. I became acquainted with several Johnny Hallydays.

Paraphrenia, in the precise case that we have just cited, shows us to what degree everything is played out at point N', where desire denies the precedence of the other's desire on which it is modeled and asserts its priority and precedence by inverting the physical chronology of events. The delusional patient is convinced that he is the true Napoleon and that it is the "other" who imitated him while, in the process, committing errors that testify to the hoax. Beginning at this nodal point N', there arises a delusion wherein the patient revises history as it is written in books.

Don Quixote

To take a similar but nonetheless different example, this one drawn from literature, in Cervantes's novel *Don Quixote* the hero is not completely paraphrenic because he does not question the very existence of Amadis of Gaul. He recognized that Amadis was the most perfect of knights and declared that he was going to try his best to resemble him in every way. Thus, at nodal point N', Don Quixote recognized the priority and precedence of Amadis de Gaul's chivalrous desire with respect to his own, which was only the copy, a copy that attempted to come as close as possible to the original.

Mimetic desire can be transmitted by the eyes or by the ears. For Don Quixote, it was by reading that he absorbed all of the stories of chivalry and lived in that world. We are witnesses, at the very beginning of Cervantes's novel, to something that we rarely observe, the search for the ideal model, for most of the time the subject is not even aware that he has taken such and such a person as a model. And we have an example of what we could call the "onset of paraphrenia," except that in this case, instead of being encysted, the delusion invades the entire field of the real and the knight, "having completely lost his mind," changes his life utterly to conform to the model of knights errant whose exploits he has just read about. He could instead have written books, and by writing have adopted with respect to his narrative the necessary distance for that narrative to remain imaginary and for him to remain in the real. But instead, Cervantes shows us how, just as in certain stories or films like Mary Poppins or *The Lion, the Witch, and the Wardrobe*, where the characters cross over into another realm, Don Quixote dove headfirst into the world of chivalry and became completely immersed in it.

◆ ◆ ◆

We must thus distinguish between paraphrenia properly speaking, where the subject claims precedence for his desire over the other's desire, and "quixotic" delusion, where the model is not denied as such. Moreover, it is certain that all the Roman emperors who were called Caesar must have taken themselves for Julius Caesar, but they did not claim to be him, but simply to be copies that were as faithful as possible.

Why Napoleon or Caesar? Because obviously, if you're going to go to the trouble to claim a desire as your own, it might as well be an exceptional and dazzling one. You will not go about claiming to have adopted the desire of someone banal. Thus paraphrenic delusions produce Napoleons, de Gaulles, Christs, and perhaps Muḥammads in certain countries, but never stable boys or garbage men.

Paraphrenia is characterized by the fact that normally paraphrenics "encyst" their delusion. Thus, for example, a baker called his bakery Duc d'Orleans Bakery, and when one asked about this in a friendly way, he recounted that he was in reality the real duc d'Orleans, descended from the authentic line of Bourbons and that the d'Orleans family that was spoken of in the newspapers were a gang of usurpers. But that did not prevent him from keeping his delusion in a bubble that enabled him to tend to his daily tasks and to be an excellent baker. Even if he was the only one to know that he was the real duc d'Orleans, that was enough to console him for the banality of his daily life.

In literature, a good example of paraphrenic delusion seems to me to be the famous *Alice in Wonderland*, by Lewis Carroll. Alice is the only one to enter Wonderland. There she encounters new models: bizarre, eccentric, fantastic, fabulous ones, but benevolent, nice, friendly. None of them is hostile, none of them is a rival. Alice, having returned from her journey, will live with her memories, but they will not prevent her from leading a normal daily life. Her memories of that marvelous trip would be the equivalent of an encysted delusion.

◆ ◆ ◆

In paraphrenia, everything is mimetic. This mimeticism is "positive" insofar as the historical personage is recognized by others, which is to say because of his fame. And it is negative in the sense that it denies the other's existence as a model and affirms that the original is in fact the deluded person, the other being nothing but a copy who takes himself for the original. This temporal and existential inversion is the very essence of paraphrenic delusion.

And yet we remain in the register of the model. The true Napoleon is not the paraphrenic's rival; he is his model, even if he is supposed to have usurped his identity.

◆   ◆   ◆

Revisiting all of psychopathology with mimetic glasses shows us two things: on the one hand, all normal, neurotic, and psychotic phenomena exist in a continuum due to the gradual exacerbation of the mimetic mechanism; on the other hand, neurotic phenomena are essentially due to a claim at nodal point N and psychotic phenomena to a claim at nodal point N', the two claims being always present, but the emphasis being placed more strongly on N in neurotic processes and on N' in psychotic processes.

You will tell me that this changes nothing. But to the contrary, this perspective changes everything, because instead of having photographs of "structures" without any relationship among one another as in French nosography, and innumerable clinical tableaux as in American nosology (DSM-V and CIM-10), we are entering a scientific space, because for each patient we explore the situation of points N and N' shared by everyone and which constitute normality if there is no problem, neurosis if the problem arises mainly at N, and psychosis if the problem arises principally at N'.

Paraphrenia thus has its source, like many psychic phenomena, in the third brain because it originates at point N'. But it is dressed in rationalizations provided by the first brain, which has historical, logical, and military arguments at its disposal to support the claim made by the third brain. Once produced and coiffed, the delusion is dressed up by taking from the second brain feelings of euphoria, satisfaction, and sometimes irritation with regard to the pseudomodel. But we have seen that, in most cases, the second brain is little called upon, to the extent that, because the delusion is encysted, it does not entail a very great affective or emotive reaction. In a word, the delusion is a delusion of the imagination and does not trigger the passions of the second brain.

◆   ◆   ◆

It can be said that the "asylums" of the past deserved their name. They were places where the "mad" sought asylum because they could not live with their delusion in the exterior world. But inside, everyone lived a "crazy" life. Even the doctors who lived there were in constant contact with the patients, and they arranged things in such a way as to benefit from each patient's talent when they were not experiencing delusions. Thus the Marquis de Sade wrote plays that were performed in the psychiatric hospital where he was locked up. I myself had a delusional manic-depressive who was a wonderful secretary at the Sainte-Anne Hospital, but I had to put up with his eccentric behavior. Moreover, a chronic delusional patient, who dressed up as a soldier from the First World War, washed the cars of all the doctors with a professional conscientiousness worthy of the highest praise.

◆ ◆ ◆

Here I would like to pause for a few reflections and general remarks. Let us summarize, even if that means I am repeating myself: psychopathology only exists against an anthropological and cultural backdrop, and thus the psychiatrist can only truly understand the nuances of his patient's thought and the various dimensions of his pathology if they both share the same culture. For example, in the United States, when I saw patients, my colleagues would say: "But of course you understand what he is saying." And I would reply: "Yes, but not what he means."

In France, I understand all cultural, historical, and literary allusions. The universal similarity of pathologies is due to the third, mimetic brain function that ensures the constancy of the mechanism (the model as . . .). But the cultural background is the cloakroom where the odds and ends that the mechanism dresses itself up in are found, even if that mechanism is identical in all cultures. That is why a neurosis like hysteria has a history and a geography: not the history and the geography of the fundamental mechanism, which operates on the level of the interdividual rapport, but a history and a geography that vary according to the justifications of the first brain function and the emotive and affective reactions of the second. Whence the question asked by Laure Murat: "Can the history of madness possibly avoid taking into account the madness of history?"[1] She mentions the fact that when the ashes of Napoleon were brought to France, the psychiatric hospitals witnessed the arrival of a crowd of emperors, which obviously shows that the ambient culture has a direct influence not on the mimetic mechanism that governs the interdividual rapport and thus not on the third brain function, but on the dress provided to this mechanism by the first brain function on the intellectual level and the second on the emotional level. That is why today people (at least in France) have a greater tendency to take themselves for a famous actor or singer or TV news anchor than for Jesus.

In the same way, in the wake of a collapse of religion and above all the Catholic religion, mystical delusions have practically disappeared, whereas they were extremely frequent in the Middle Ages. Obviously, if for example someone believes in the phenomenon of possession in the context of a traditional African animist culture and declares that he is possessed, that does not in the least make him delusional, whereas if a Parisian declares that he is possessed by Joan of Arc, he is delusional, because French culture is not prepared for phenomena of possession and is not equipped to handle them.

It thus appears essential to me to seek to discover the universal springs of psychopathology, which are located in the mimetic brain function. The existence of this third brain can only be established by means of the research I mentioned above, that of Girard, Gallese, Meltzoff, and even Damasio and all those who have discovered the importance of emotional intelligence. It makes it possible for us to see, across historical, geographical, and cultural variations, the unity of the mechanism that underlies them.

For example, with respect to what was said about paraphrenia, if a French person seeks to take the model as such and to construct a delusion on this mechanism, he will obviously opt for Napoleon, de Gaulle, Gérard Depardieu, and so forth. But if an Indian becomes paraphrenic, he will take Gandhi, Nehru, or Buddha as a model. Personally, I have never seen a French patient take himself for a celebrity from another country—for example, Obama or Kennedy. By the same token, it would surprise me to find an English gentleman who took himself for Napoleon.

◆   ◆   ◆

Until now, psychopathology provided photographs of illnesses, but there was no rationality and thus no possible scientific approach making it possible to explain that so-and-so took himself for Napoleon, that another was possessed by a spirit, and that a third was persecuted by the CIA or the KGB, and so on and so forth. With the hypothesis of the third brain, the mimetic brain, which I am trying to develop here, we understand how a basic mechanism can of necessity produce an intellectual and emotional "dressing" that justifies it and gives it all of its historical, geographical, cultural, and anthropological colors. The three possibilities of the interdividual rapport are always the same: the model as such, the model as rival, the model as obstacle. All psychopathology can be expressed by the overlapping of these three forms with the three possibilities that exist in each person's memory, that is to say in each person's psychological time.

1. If psychological time and memory are content to forget physical time and the unfolding of the phenomena that have created the self, we find ourselves in a "normal" structure.
2. If, to the contrary, memory is fixed on point N to claim ownership of its desire, we find ourselves in a neurotic structure, and this claim will be manifested in very different ways according to whether the model is seen as a rival or as an obstacle.
3. Finally, if memory seeks to turn back physical time at point N', this means that desire is claiming its anteriority/precedence with respect to the other's desire, which engendered it, such that the two desires compete with each other in a sense upstream of the patient's self's existence so that the patient no longer sees the difference between himself and his model, himself and his rival, himself and his obstacle. In physical time, desire produces the self, but if we place ourselves before the creation of the self, desire does not see the difference between it and the other's desire because the self capable of perceiving this difference has not yet been created.

The Other as a Rival

When the model is a rival, suspicion infiltrates the first brain. If all models are experienced as rivals by the third brain, the first brain sees enemies everywhere, or to the contrary concentrates all of its rivalry on one suspected enemy or rival: the enemy is not suspected because I know he has something against me; rather, I am sure he has something against me, merely because I am suspicious of him.[2] Thus, for example, Othello suspects his supposed rival so strongly of having seduced Desdemona that he wants to deprive him of the possession of the desired object. Being unable to get at the rival himself for the simple reason that he does not exist, Othello ends up adopting the delusional attitude that consists in killing Desdemona in order to radically deprive all potential rivals of her.

In daily clinical psychopathology, paranoia consists in being persecuted by a designated rival and paranoiacs react to that persecution by legal or more rarely by aggressive means.

In other types of delusion (chronic hallucinatory psychosis), the persecution is actualized by the first brain in the form of hallucinations. In other words, by an otherization of thought and perception, the patient truly hears his enemies insulting him and calling him, for example, "fairy" or (if the patient is a woman) "whore," or some other name. Out of this delusional and hallucinatory experience, the subject will construct a story making it possible to explain what he is going through. This story brings in forces like the CIA or the KGB with sophisticated means and able to send waves through walls and ceilings. Here we have a new example of the cultural dressing of delusion since in the Middle Ages sensations and hallucinations were ascribed to demons and witches.

The claim here is made at N', in the sense that desire frenetically claims precedence over the other's desire, and the claim is expressed by a phrase that is uttered by all these persecuted patients: "I am being followed." That is to say my rival is pursuing me, he is behind me, he is lagging behind, which of course means that I am preceding him, that my desire is anterior to his.

Interpretative, paranoiac delusions use all the elements of a given situation to further this claim. A lady leaves her building and sees, on the other side of the street, a car parked with a woman at the wheel. At the moment the lady comes out, the woman picks up her portable phone. The lady immediately goes back up to her apartment and lays into her husband, who does not understand what is happening to him. She says to him: "I am sure that woman is your mistress and that she was phoning you." And from there, the more he comes up with logical arguments to disprove her theory, the deeper he sinks. Once you engage with delusional logic, every argument gives rise to a counterargument. For example, if the man says: "Calm down, I tell you it's nothing," she will reply: "You see? You are embarrassed." And if he says: "No, I'm just embarrassed to see you in this state," she will retort, triumphant: "Yes, you are embarrassed because you have a mistress!" And so on and so forth.

Emil Kraepelin defined paranoia as a "relational delusion" and Ernst Kretschmer would emphasize this point even more, which coincides completely with my approach. It is indeed the relationship with the other that is deformed and distorted, and Lacan specifies that this relational delusion "describes the multiple subversions provided by the ill patient to the meaning of the gestures, words, trivial facts, as well as the spectacles, forms, and symbols that he apprehends in daily life."[3] It is typically at the level of the third brain—the relational and mimetic brain—that the disorder is situated, and from there springs the delusional interpretation of the rapport with the other, which will be taken up and rationalized by the first brain function in two opposite ways: the "delusion of persecution," which recounts and rationalizes a persecution, a jealousy, hypochondria or a legal claim; delusions of grandeur, which produce "inventors," mystics, and erotomaniacs.

◆　◆　◆

Here I would like to open a parenthesis to go back to Lacan, who tells us: "The consensus among psychiatrists, we know, ascribes the genesis of the illness to a progressive personality disorder."[4] It is of course paranoia that Lacan is talking about. If I cite the reference to "a progressive personality disorder," this is because I am in complete agreement with Lacan in what he says later on, namely that in fact "rather than a personality," it is a matter of "a succession of personalities."[5] And he goes on: "Are these not the very transformations that, according to the case, we call enrichment or abandonment of ourselves, progress or conversion?"[6]

Here there is cause for optimism: if the successive personalities can become more and more sick, constructed on increasingly rival interpretations of the interdividual rapport, they can also become less and less adversarial, evolving in the direction of progress.

And Lacan poses the question that I have been asking ever since *The Puppet of Desire*: "What remains of our continuity?"[7] However, he distances himself from my approach at the moment he is closest. He finds it too easy to "conceive the person as the link that is always ready to be broken—and which is besides arbitrary—of a succession of states of consciousness, and to base his theoretical consideration on a *purely conventional self*."[8]

The "conventional self" is what I call the "self-of-desire," coproduced by the model's desire, which by means of suggestion elicits the imitating desire, which will forge and form the self. The "normal" self and personality are thus indeed conventional, that is to say consensual, which means that in "normal" people the same gestures, the same signs, the same events have the same signification and are thus the object of a shared understanding. Lacan affirms this himself when he writes that if "personality has a certain unity," it is that of a "regular and comprehensible development."[9]

Mental illness thus comes from a break in the harmony of the mimetic brain: when the interdividual rapport becomes rivalrous, from the persecuted, jealous, or erotomaniac paranoiac's point of view, this leads the first brain to rationalize and justify its rivalry by constructing a delusional reasoning process, that is to say one that has become incomprehensible for others. This incomprehensibility aggravates the process and separates the rivals more and more while exacerbating the rivalry. The exacerbated rivalry is reverberated by the second brain function as a passionate exaltation that presents itself as a manic excitation, that is to say a mood disorder in the sense of hyperthymia. All the authors agree on this point, and it is by drawing on Lasègue, Legrand du Saulle, Falret, Köppen, and Sérieux and Capgras that Lacan writes: "Manic exaltation is part of the classical clinical tableau of persecuted persecutors."[10]

It is equally clear that the failure of the claims, of the legal procedures, of erotomaniacs' attempts to make contact with their idol, in sum of the opposition of reality to the triumph of their conflict, often leads to depressive and even melancholic states. Here we can see the dialectic of the three brain functions emerging: the third is involved in rivalry and the rejection of the other, of otherness, the first rationalizes this rejection and finds legal, ethical, moral, political, and religious justifications, and the second accompanies the rivalrous passion as it "goes to war" in the enthusiasm of its delusional certitude, but translates the combat's failure by a state of depression.

Another way of experiencing that failure consists in seeing the rival transformed into an obstacle against which the self will be shattered, as we will see later on, but Lacan points out that in "certain cases . . . fleeting or durable schizophrenic manifestations are detectable in the sick patient."[11] The evolution of the interdividual rapport makes it possible to "jump to a new square" in our table (see chapter 9) showing the overview of my proposed nosology, and shows once again that it is the interdividual rapport that determines pathology according to whether it is in the model-model, model-rival, or model-obstacle position. With respect to this slippage, Jacques Lacan reports an observation of Professor Claude in 1925 of "a very beautiful case where a known paranoiac psychosis, which for a long time had been compatible with an effective professional life, although *rich in conflicts*, evolved toward a paranoid psychosis."[12] Little by little, the model had ceased to be a rival and had evolved into an obstacle.

Finally, let us cite Lacan once more to confirm our conception according to which the delusion bears witness to a relational problem, and psychological or psychopathological reality is always situated *between*: "If you take a close look, you see that the symptoms are not manifested with respect to any old perceptions—of inanimate objects without affective signification, for example[13]—but most especially with respect to *relations of a social nature* with the family, colleagues, neighbors. Similarly, the reading of the morning paper . . . which is the sign of the union with a larger social group. The interpretative delusion . . . is a delusion of the doorstep, the street, the forum."[14]

◆　◆　◆

I would now like to read a few passages written by Jean-Jacques Rousseau, whom Lacan calls a "paranoiac of genius,"[15] which in my opinion mark out the philosopher's trajectory, leading him to sink into paranoia as a result of thoroughgoing misrecognition of mimetic realities. We will see that it is the constant negation or misrecognition of otherness, which classical psychiatric authors call the "overestimation of the self," that is at work here.

For Rousseau, men initially become wicked not because they imitate one another and compare themselves to one another but because of private property, which is to say that, for him, the source of evil is in the object. Here are the first lines of the second part of *Discourse on Inequality*: "The first man who, having enclosed a piece of land, thought of saying, 'This is mine' and found people simple enough to believe him, was the true founder of civil society. How many crimes, wars, murders; how much misery and horror the human race would have been spared if someone had pulled up the stakes and filled in the ditch and cried out to his fellow man, 'You are lost if you forget that the fruits of the earth belong to everyone and that the earth itself belongs to no one!'"[16]

In his dialogues, Rousseau diagnoses the evil that he calls "amour-propre," or "self-regard," as the concentration on the rival and on rivalry rather than on the object to be attained. By saying that, he is very close to our way of thinking because indeed, exacerbated rivalry leads the rivals to concentrate on each other and lose sight of the object that first triggered their conflict. The problem that takes us far from Rousseau's way of thinking is that he believes there is a state of nature in which no such desires and no such rivalries exist. And that state of nature, as we just said above, was, according to him, destroyed by private property.

Rousseau makes any and all differentiating principles, which is to say difference itself, responsible for wickedness and violence. He asserts that it is difference that produces rivalry, when in fact it is rivalry that produces difference: my neighbor's property is bigger than mine; he is thus my rival, but it is not the size of his property that makes him my rival, it is jealousy, that is to say my mimetic desire, because the property that I have myself could be enough for me if I didn't compare it to my neighbor's but instead to the still smaller property owned by another neighbor. Comparison can be a source of frustration or exaltation depending on whether it bears on what one has more or less of compared to others. More generally speaking, we are always someone's rich person and someone else's poor person.

Let us linger a little over this famous *Discourse*. In the first part, Rousseau's bucolic vision appears very romantic: "The earth, left to its natural fertility, and covered with immense forests that no axe had ever mutilated, would afford on all sides storehouses and places of shelter to animals of every species."[17] One would think oneself at the Petit Trianon with Marie-Antoinette playing with her sheep and cows! The only interesting point is that Rousseau says that man *imitates* all the animals and appropriates each one's art: "Man, dispersed among the beasts, would observe and imitate their activities."[18] Obviously, this is a total misunderstanding of the animal world—we cannot imitate animals because they have a sense of smell that is infinitely superior to our own, a more piercing sight, a more sensitive hearing, not to mention the fact that they run faster than we do, and so forth.

A little later comes the outline of a sort of Darwinism. Accustomed to the rigors of the seasons, to experiencing fatigue, and to defending themselves against other ferocious beats, "or to escape such beasts by running faster, men develop a robust and unvarying temperament."[19]

Those that didn't die of cold, and who ran faster than the wild beasts, must not have been very numerous. But here we already see the dramatic

conflict of violence emerging. Men do not (yet) attack one another mutually, but they have to defend themselves against ferocious beasts. In passing,
Rousseau critiques Hobbes, who in his view has too high an opinion of man's
naturally intrepid nature. Man does not, Rousseau claims, attack and fight at
every opportunity, as Hobbes would have us believe. To the contrary, he is,
in the "state of nature," extremely timid, always "trembling," and ready to flee
at the slightest sound.

He buttresses these assertions with references to some mostly forgotten
philosophers (Cumberland and Pufendorf). But of course all of these philosophers have illusions that are either too positive or too negative. The answer
was provided by Hans Selye in his studies on stress,[20] where he teaches us
that stress can trigger one of two reactions described by the various authors
cited: fight or flight. The choice between the two is a function not only of the
intrinsic character of the human being in question but also of the objective
analysis of the situation: the decision to flee is directly proportional to the
size or number of the ferocious animals that are attacking you; the decision
to fight is made when you think you have a good chance of winning. But
Rousseau does not see any of this.

He who introduced imitation earlier, without paying it the attention
it deserves, now speaks of *comparison*, without perceiving its importance
either: man lives among animals constantly, and therefore begins to "measure
himself against them" and "soon makes the comparison"[21] that he surpasses
the animals in skill just as they surpass him in strength—and in this way he
ceases to fear them. Here, the comparison with animals, at once too optimistic and unrealistic, appears to be always in favor of man, which is more than
improbable. Moreover, Rousseau has not yet arrived at the idea that men can
compare themselves to and imitate *one another*.

And yet mimetic contagion is present somewhere behind the Rous-
seauist illusion, as a possibility that he never speaks about but that nonetheless exists. Man must contend with natural infirmities, the weakness of
childhood and of old age, "and illnesses of every kind, melancholy proofs
of our own weakness, the first two being common to all animals, the last
belonging chiefly to man as he lives in society."[22]

If illnesses occur only to human beings living in society, it is obviously a
question of epidemics. And the notion of *contagion* is yet another mimetic
notion that Rousseau mentions, addressing only its negative side, that is to

say the transmission of illness from one human to another, neglecting the fact that knowledge, learning, and language (which Rousseau doesn't mention for the moment and which constitute a true difference between humans and animals) can also be transmitted from one human to another. For the third time, Rousseau bypasses mimetic desire while at the same time evoking its aspects. He observes imitation, comparison, and contagion, and he even has the intuition of a victimization of the weakest and of a sort of generalized Darwinism giving rise to the selection of the healthiest animals and the most resistant humans, but without establishing a link among these various phenomena.

Later, he speaks of domestic animals. For Rousseau, domestication both humanizes and denatures or "bastardizes" animals, for this time they are the ones who imitate humans instead of being imitated by them in a state of nature. Curiously, Rousseau has the prescience to point out something like emotional intelligence. He sees clearly that cognitive intelligence, if it is not completed by emotional intelligence, which he calls "passion" or "desire," is ineffective: "human understanding owes much to the passions, which, by common consent, also owe much to it."[23] It is by experiencing passion, says Rousseau, that the human ability to reason is perfected. We only seek out understanding because we want to experience pleasure, and if we didn't have fears and desires, we would never even go to the trouble of reasoning.

It is interesting to see how, for Rousseau, desires and needs are, in the "savage" human being, originally confused. Here again, he comes very close to the idea that desires copied from other desires no longer correspond to needs. From this point of view, mimeticism could be considered as corrupting the state of nature, if that state existed, if man were not, as Aristotle said, a "political animal," which is to say incapable of living except in society.

It is because Rousseau does not see or denies mimeticism that he becomes paranoiac in my opinion: to deny mimetic desire is to deny otherness. The pathological isolation that refutes mimeticism by adopting a pose of inimitability appears in Rousseau's later *Reveries of a Solitary Walker*, which are in reality the complaints of a scapegoat. To declare that one is inimitable is to declare that others are seeking to imitate you. If they are seeking to imitate you, that means that their desire must tend to copy yours. And if their desire copies yours, this is obviously because yours comes first. You are in the lead, alone, preceding all the others and thus followed, even pursued, by them.

In the *Reveries*, it almost looks like Rousseau is doing field tests of Girardian psychopathology! The paranoiac has nothing but enemies. He is alone on earth, he is the perfect scapegoat, at least in his own eyes. And in his first "walk" Rousseau underlines his solitude: "So now I am alone in the world, with no brother, neighbor or friend, nor any company left me but my own. The most sociable and loving of men has with one accord been cast out by all the rest."[24]

The unanimity posited by René Girard as essential to the scapegoat mechanism is underlined by Rousseau. But in him this unanimity is subjective because it issues from his paranoia and not from a real mechanism of social persecution. It is in fact Rousseau who excluded and expelled himself, and not society that rejected him, as he claims: "Could I, in my right mind, suppose that I, the very same man who I was then and am still today, would be taken beyond all doubt for a monster, a poisoner, an assassin, that I would become the horror of the human race, the laughing-stock of the rabble, that all the recognition I would receive from passersby would be to be spat upon, and that an entire generation would of one accord take pleasure in burying me alive?"[25]

Such is the scapegoat's complaint, and you would think you were hearing Oedipus's complaints on the road to Colonus; such is the paranoiac's complaint, the difference being that paranoia is the loudspeaker for the scapegoat to which it gives voice. In primitive religions, it is the myth and thus the "lynch mob" that speaks for the scapegoat, but only after its death. And after that death, the divinized scapegoat makes recommendations, prescriptions, injunctions, prohibitions, and brings peace—all of that, of course, without his knowing it, since he is dead. Rousseau, for his part, gives nothing at all to his "persecutors" because he has excluded himself while still alive. What is interesting about his case is that he describes the experience of the scapegoat, that is to say the feeling of the psychotic who sees enemies everywhere, while considering that he is the victim of injustice because he is superior to all who persecute him.

Moreover, it is quite interesting to see that reality (in the negative sense) catches up with Rousseau in his *Letter to M. d'Alembert*. Here he identifies in a certain sense with the misanthrope in Molière's famous play. Now the misanthrope, like Rousseau himself, is caught in the mimetic trap. He is in rivalry with everyone and wants to distinguish himself from others at all costs: "I spurn the easy tribute of a heart / Which will not set the worthy man apart" (I, 1).[26] Which means that he wants to be distinguished and advantageously compared, and ultimately recognized as unique and thus as superior to all other human beings: "I choose, sir, to be chosen; and in fine, / The friend of mankind is no friend of mine" (I, 1).[27] As a result, wanting to be superior is incompatible with the fact of being "the friend of mankind." The paranoiac and the misanthrope are, in the final analysis, one and the same.

In the same play, Célimène admirably describes the semiology and nosology of the paranoiac/misanthrope:

> You don't expect him to agree with us,
> When there's an opportunity to express
> His heaven-sent spirit of contrariness?
> What other people think, he can't abide;
> Whatever they say, he's on the other side
> . . . .
> Indeed, he's so in love with contradiction,
> He'll turn against his most profound conviction
> And with a furious eloquence deplore it,
> If only someone else is speaking for it. (II, 5)[28]

This is a perfect illustration of the mimetic rivalry that consists in asserting that my feelings, my ideas, and ultimately my being are superior and of course anterior to those of other people. And if I catch one of my ideas on someone else's lips or one of my feelings in his heart, I immediately reject that idea or that feeling, for I consider that it cannot belong to me, simply because, inasmuch as another has it, it has become contemptible and unworthy.

Paranoia is the psychotic expression of rivalry.

The Other as an Obstacle

When the model evolves into an obstacle, the obstacle-desire will ultimately forbid not the possession of an object but the very being and existence of the imitating self. Shattering against the obstacle (whence the constant clinical observation of a fragmented self), the self explodes into a thousand pieces like a broken mirror, the shards of which can still reflect a part of the environment, although the whole picture can no longer be reflected, apart from a few recurrent elements.

The first of these elements is called precisely the "sign of the mirror": schizophrenics look into a mirror and no longer recognize themselves. This symptom confirms the interdiction of being, which is pronounced by the obstacle, the mirror's message being: "This is not you and therefore you do not exist."

A variant of the sign of the mirror is dysmorphophobia (also known as body dysmorphic disorder), which consists, for the schizophrenic, in seeing a part of his face differently than it really looks and holding it responsible for all his problems: if he has a nose job, for example and the bump disappears, he thinks that he will no longer have any problems. This symptom is the habitual trap that plastic surgeons must avoid.

Thus, this pathognomonic symptom of schizophrenia (that is to say a symptom whose presence guides the diagnosis toward this psychosis) involves two elements:

1. This nose (or this scar, or my ears that stick out, and so on) is bothering me, and all my problems are due to it. It is not mine. It is not me and I do not recognize it as such. (This part of the body has been, so to speak, "otherized.")
2. This nose thus represents the other as obstacle, otherness "obstacle-ized." It is indeed the only obstacle to the fulfillment of my desires: seducing men or women, for example. All I have to do is cut it off, eliminate it, to get rid of the obstacle and clear the way for my desire to be realized, the psychotic thinks.

Alas, a surgical operation, even one that is successful, will only displace the problem and the model-obstacle will then be represented in another form or by another body part or by claiming that the surgeon has botched the operation.

A second sign: the schizophrenic hears one or several voices that insult him, or comment on his actions, or repeat his thoughts, or criticize his actions. This is what Clérambault called the "syndrome of mental automatism," which we already encountered in the chronic hallucinatory psychosis.

It is very interesting to see that if the hysteric represents the model-rival by otherizing a part of the body, the schizophrenic represents the model-obstacle by otherizing a part of his thought or sensation, in other words by attributing the voices that he is hearing or the sensations that he is feeling to the action of another, who is external to him.

Obviously, in hysteria as in schizophrenia, the model-rival and the model-obstacle are real, but their otherness is not recognized, and it is on this otherness that pathogenic misrecognition bears. At N', it seems that when desire rebounds off the obstacle-desire or the obstacle-desires with which it is confronted, the self no longer knows how to, or no longer wants to, or can no longer distinguish if the desire that rebounds is the other's or his own. This would explain the subject's experiences of influence, of echoed thoughts and mental automatism. By the same token, the words and thoughts are repeated, which is consistent with the rebounding, and the actions are commented on in a sort of affirmation of the precedence of the self's movement with respect to the other's, which can only offer a posteriori commentary. For example, the patient is reading and he hears a voice that says to him: "You are reading . . ."

The doubling back of desire on itself, after having rebounded off the obstacle-desire, is expressed by the experience of influence that Henri Ey describes in the following manner: "The sick patient is subjected to a series of communications, figurative break-ins, or remote-controlled thought processes. Someone is reading his thoughts, trying to steal them, or imposing those thoughts on him. Fluids, waves, and radar track his thoughts and control them. This experience is generally associated with a more or less rich context of acoustico-verbal, sensory, and psychomotor hallucinations. When he gets close to such and such a person or object, he feels the fluid or else hears words being spoken."[29]

The lack of affectivity, the incapacity to dissociate the symbol from its referent—for example, "I was told to take the plunge," and the patient jumps into the Seine, or else, "I'm the boss here, get that into your heads," and the axe falls on the skulls of the two parents who were having breakfast with a young schizophrenic—all of this bears witness to the fundamental incapacity of an imitation whose essence is to put oneself in the other's place. These patients cannot put themselves in the other's place. The other comes back at them like a boomerang whenever it appears.

The reality is that, unable to imitate others but nonetheless caught in the mimetic network that is social reality, the schizophrenic senses through various metaphors that others are imitating him (he is no more mistaken than was Mesmer, who hypostasized the interdividual rapport in the form of a physical and astral fluid). Just as the paranoiac asserts the preexistence of his desire with respect to the other's rival desire, so does the schizophrenic bump into the obstacle formed by the other's desire, which in a way is but his own rebounding off the other's and coming back otherized. By rebounding off the obstacle, his own thought comes back to him as if it had been deciphered by the other. The schizophrenic otherizes his thought or his desire in the form of external voices to represent the relationship to the other, the model-obstacle, who is at times a powerful and insurmountable persecutor, and at times an obstacle that must be destroyed or killed. The young patient who kills his two parents at breakfast thus carries out the orders of "his voices" when he swings into action, which is indeed the major risk in schizophrenias.

A young patient told me one day: "When my mother came to see me, I heard a voice that was telling me: 'Kill her!'"

Instead of denying the hallucination, and telling the patient that it was impossible that she heard voices and that we were going to cure her and get rid of them, I said to this young patient, whom I knew hated her sister-in-law bitterly: "And when your sister-in-law comes to see you, do you hear the voice saying to you, 'Kill her!'?"

"No," she replied, surprised.

"How strange, did the voice mix up the adversaries? I am sure that if you had the choice, you would rather kill your sister-in-law than your mother, no?"

This absurd remark made her laugh, for the first time in a long time. Laughter bears witness to comprehension, an infraverbal understanding. This laughter showed that the interdividual rapport between her and me was not stuck in the "obstacle" position. I think that this was a small step forward in the treatment of that patient. We had established a sort of complicity, integrating the hallucinatory phenomenon.

◆　　◆　　◆

It seems to me that all the classical authors have always distinguished various clinical forms of schizophrenia. In light of our reflections up till now, I think that we can distinguish between two major varieties:

1. A first group including hebephrenia and hebephreno-catatonia, which seem to be organic or somatic or neurological forms of schizophrenia.
2. Schizophrenias that could be qualified as "psychogenic," including what could be called "simple or paranoid schizophrenias," where a major role is played by the model taken as an obstacle in the interdividual rapport.

To illustrate this second variety of schizophrenia, of psychogenic origin, I will cite the case of two brothers. The older brother amassed a colossal fortune and became the head of the family, but he never married. On the other hand, the younger brother married and had two children, including an older son. When the son arrived at adolescence, he performed rather brilliantly at school, but gradually realized that his father was hardly brilliant, that he (the father) was supported financially by his brother and that, as a result, the father was not a worthy model with whom he could identify.

The son became more and more fascinated by his uncle, whom he obviously admired, like the whole family, and he ended up offering to work for him. But the uncle wouldn't hear of it. Little by little, the young man latched onto the idea that this absolute and ideal model should have been his father, but faced with the radical impossibility of that idea and confronted with the rejection of his uncle, who did not wish to "adopt" him, he started to become delusional in a fluctuating way. He was persecuted by his superiors, the victim of all kinds of persecutions, and he was fired from all the jobs he took, and soon he withdrew into himself, his personality became fragmented, split into pieces, and he was no longer able to put it together and thus to act.

In this case, taking the model as a model was impossible and taking him as a rival unthinkable, and so the obstacle towered in the young man's path, a goal that was impossible to attain, an objective that was out of reach, a model who was not only inimitable but who was rejecting him.

He said to his uncle: "Come take me away, kidnap me. When you are on the road to the airport, I will be at a certain intersection, the car will slow down, I will jump in, I'll get in to your private jet with you, and I will be healed because I will be with you, everything will be okay."

Obviously the uncle refused, because he saw clearly that the young man was deranged, and he didn't want to take that responsibility or put himself in an awkward position vis-à-vis his brother.

What would have happened if the uncle had given in to the young man's request? In my opinion, the young man would have been very happy. He would have worked with his uncle, and very soon he would have taken him as a rival, wanting to show him that he could do things better than him. He would thus have "jumped to another square" in our clinical tableau, but the absurdity of the undertaking would have led to a rivalrous psychosis. He would have begun to give orders at the office and to tell everyone, for example: "My uncle said that it's black? No! I'm telling you it's white!" A harmonious relationship cannot be constructed on the rejection of one's own father. Thus, in a certain manner, the worm was in the fruit from the beginning.

Here we are touching on the sometimes oscillating kinship between paranoid psychosis and paranoiac psychosis. The kinship and difference between these psychopathologies have been noted and underscored by European psychiatry.

In the young man's case, we can see clearly that the model is loved, that initially he was a model and only becomes an obstacle gradually, inasmuch as he refuses the nephew's imitation and identification. The latter's desire, copied from and inspired by the model's desire, is at the same time rejected by him, forbidden from existing by the very person who engendered it. It is in this sense that the young man shatters himself on the obstacle and that the self explodes in the collision. But the second brain still has affection for the uncle, and that affection enters into conflict with the resentment and hatred generated by the uncle's "rejection," by his metamorphosis into an obstacle. That is why the schizophrenic *simultaneously* experiences two contradictory feelings: love and hatred. This is what psychiatry has always called "ambivalence," which is a major symptom of schizophrenia.

The uncle being the absolute model, the self that is formed in the nephew by imitating that model is a divided self, the self-of-two-contradictory-desires: "Do as I do, be a hardworking, responsible man, and at the same time I don't want you to be like me, a powerful billionaire. I refuse that desire." The model's contradictory injunction probably corresponds to what Gregory Bateson envisaged as being the mechanism that triggers schizophrenia, under the name "double bind."

◆ ◆ ◆

It seems to me that in Nietzsche's letters to several of his correspondents, it is possible to discern the philosopher's advance toward a delusion of the schizophrenic type, as Nietzsche gradually shatters himself against the formidable obstacle that Wagner constitutes for him.

In the first letters, Nietzsche recognizes or in any event declares that Wagner is his master and his model just as much as Schopenhauer is: "Truly the best and loftiest moments of my life are linked with your name."[30] He even goes so far as to speak of *religione quadam*, that is to say a practically religious veneration. In a letter of January 1872, he affirms that every page of the essay he is in the process of writing is an attempt to thank and recognize Wagner for everything that he owes him and that heretofore, whenever mention is made of him, it will be in connection with Wagner: "My most revered Master, at long last my New Year's greeting, and a Christmas offering."[31] In mid-November 1872, a little problem begins to rear its head: he doesn't have any students. And in April 1873, he acknowledges his inferiority: "I learn and perceive things very slowly, yet every moment that I am with you I meet something I've never thought of before and yearn to make mine."[32] In this letter to Wagner, he recognizes that the composer is his absolute model and that his desire is to imitate him.

Five years later, in January 1878, he has just received *Parsifal*, and in a letter to Reinhart von Seydlitz, the criticism begins: "It's all too Christian, dated, narrow-minded. All sorts of bizarre psychology."[33] Nietzsche has already almost reached the limit of what he can stand from this model who is going to be transformed beneath our eyes into a rival, and then into an obstacle. A later letter to von Seydlitz underlines Nietzsche's growing ambivalence: "His aspirations and mine keep drawing apart. . . . Besides, if he were aware of everything inside me that militates against his art and his aims, he'd think me one of his worst enemies—which as you know I'm not."[34] In July 1878, in a letter to Mathilde Maier, he confesses: "A baroque art full of overexcitement and glorified extravagance—I mean Wagner's: it was these two things that finally made me ill."[35] Seeking a counterweight to the Wagnerian model, he clings to the Greeks: he declares himself "immeasurably nearer the Greeks than before,"[36] and he begins, in the same letter to Mathilde Maier, to seek the means of salvation by dissociating wisdom in general from the wise men whom one transforms into idols ("before I only idolized wise men"). The psychological mechanism that is appearing consists in separating one part of the person from the other: the wise man is not wisdom. Wisdom cannot be reduced to the wise man who possesses it; it is independent from him in a certain sense. Little by little, this nuance will enable him to continue to be a philosopher without necessarily having to love or admire the sage who has been transformed into an idol: "You see, I've reached a level of honesty where I can endure only the purest of human relations,"[37] and he reiterates his attempt at dissociation when he says that one must be at once the model and the disciple: "Let everyone be his (and her) own true follower."[38] The process of dissociation unfolding here seems to herald the onset of schizophrenia.

Rivalry at N' is triggered when, in September 1878, war is declared between the two men. Nietzsche writes to Franz Overbeck: "I've finished reading Wagner's extremely bitter and unfortunate polemic against me."[39] At the same moment, at the end of the summer of 1878, he writes to Carl Fuchs that he no longer feels alone against Wagner, for the "Herr Doktor" shares his negative feelings with regard to the composer: "So you too, my dear Doctor, are having a Wagner crisis!"[40]

Finally he begins to say that his desire precedes Wagner's. It is July 1882. He writes to Peter Gast that he has rediscovered his old compositions, that he has played them and that he is amazed at how similar he and Wagner are. At nodal point N' desire discovers with surprise its similarity with that of the model and begins to affirm its priority with respect to it: "I must confess it came as a real shock to realize once again how very much alike Wagner and I are. . . . Of course you understand, dear friend, no praise for *Parsifal* is intended. What decadence!"[41] Nietzsche suggests that Wagner's talent seems to have declined since the days when they were on friendly terms. In sum, at nodal point N', he asserts that his desire preceded Wagner's and produced music of great quality and that now that Wagner is no longer, in a certain sense, imitating him, he has degenerated to the point of writing *Parsifal*.

Because Nietzsche is, after all, a genius, he begins to recognize, in a letter addressed to Paul Rée and Lou Andreas-Salomé, not only that he is sick but that moreover his sickness is in reality a narcissistic wound, a wound to his pride. He was unable to "digest" *Parsifal*. Ambivalence, another main symptom of schizophrenia, appears. Having listened to the prelude to the work, he expresses his admiration for Wagner's genius, which he compares to Dante's.[42]

But he experiences a relapse in December 1887 in a letter to Carl Fuchs: "I now regard my having been a Wagnerian as eccentric."[43] Nietzsche adds that it was a "highly dangerous" experiment and concludes: "Now that I know it didn't ruin me, I also know what significance it had for me."[44] He thinks that he has escaped Wagner's pathological or pathogenic influence, or he tries to convince himself that he has escaped it.

In a text dating from February 1888, he writes that Wagner and Schopenhauer are "as much my kin as they are my antagonists,"[45] in other words that they are as much his models as they are his rivals. And he adds that he was the first to detect the unity between the two. He is thus superior, he has now understood everything, and from the heights of his grandeur he gives a diagnosis. One easily perceives that his genius is trying to rise above and compensate for his mental illness: he sees that Schopenhauer and Wagner are models who have slipped toward rivalry.

Later his resemblance to Wagner is taken up again: he states that Wagner and himself have a sort of genius that produces the same effect. Speaking of the dying Baudelaire, and of an unpublished letter written by Wagner, he writes: "During the last period of Baudelaire's life, when he was half insane and slowly going to pieces, they fed him Wagner's music as *medicine*; and at the mere mention of Wagner's name, '*il a souri d'allégresse*' ["he smiled with joy"]. . . . (If I'm not completely deceived, Wagner wrote a letter of this sort, full of gratitude and even enthusiasm only one other time: after he received *The Birth of Tragedy*.)." Nietzsche compares the therapeutic effects of Wagner on Baudelaire to his own relationship with the composer. Whereas before there was the challenge at N' with regard to music and the priority of Nietzsche's compositions over Wagner's, now there is the fact that he and Wagner, each in his domain, have the same miraculously therapeutic effect.

But this equality will not last, because in August 1888 he condemns Wagner's music, and what I have called the "infernal seesaw" oscillates faster and faster. Nietzsche and Wagner are rivals, and it is as if they are on a seesaw: Nietzsche cannot feel superior, cannot rise in his own eyes, unless Wagner descends. In a letter to Burckhardt, he says that Wagner's operas are everywhere, that they are invading everything. In Autumn 1888 he explains to Burckhardt that now Wagner has such notoriety that even the German emperor considers him to have a national importance. "I think I have a right to do some plain speaking about this 'Wagner Case'—perhaps even a duty."[46] In other words, things must be brought back down to their true size: Wagner's fame is completely overblown! In September 1888, he declares that nobody in Germany writes like he does.

In autumn 1888, his isolation opens the door to depression. He writes to Georg Brandes to tell him that he was the only one to react favorably or even to react at all to his attack on Wagner: "For no one writes to me."[47] He has nobody gathered around him anymore. In October 1888, he writes once again to Brandes and he tries to cure himself. He announces to him that he has begun to write *Ecce Homo* and declares that this book will be an audacious discussion with himself about his own writings. In November 1888, he specifies to Franz Overbeck what he expects from this book of reflections about himself that should make him much more visible. He wants to be seen. Because the Other, Wagner, is so visible, he wants to be also.

In December 1888, in a letter to Carl Fuchs, delusion begins to set in: "Since the old God has abdicated, *I* shall rule from now on."[48] He is speaking at once of his own god, Wagner, and of that of the common run of humanity, that is to say God. God is dead, or in any event has abdicated, and thus it is up to Nietzsche to take command. But at once he comes back to saying that he would like a crowd, lots of people, talented musicians, to join him in forming an anti-Wagnerian group hostile to the Bayreuth mob. He suggests that a little brochure reporting all sorts of new and crucial things about him, and emphasizing in particular his relationship to music, be published.

In December 1888 he writes to his mother, Franziska: "As a matter of fact, your poor old offspring is now a mighty famous man. Not especially in Germany, since the Germans are too stupid and vulgar for the grandeur of my genius and have always made themselves ridiculous where I am concerned,—but everywhere else. I have only the most select admirers, highly placed and influential people, in Saint Petersburg, Paris, Stockholm, Vienna, New York. . . . No name is treated with as much distinction and reverence as mine today."[49] He has definitively rid the world stage of all rivals and now he is alone. He revels in this situation once again in a December 1888 letter to Meta von Salis, in which he declares: "I'm beginning, in an utterly incredible way, to become famous."[50]

A letter to Peter Gast—one of the famous last fragments—dated January 1889 testifies at once to Nietzsche's euphoria and to his delusion: "Sing me a new song: the world is transfigured and all the heavens are joyous.—The Crucified One." From this moment onward, Nietzsche, in his sick mind, has replaced God, has replaced Wagner, and declares himself an even greater musician, and an even more famous man, than the latter. But since God was crucified, he signs the letter "the Crucified One." He confuses the two poles because it is difficult to separate sacred kingship from the collective immolation of a victim, and in a Judeo-Christian society, it is difficult to separate divinization from crucifixion.

And yet, he tries, for in a letter to Cosima he writes: "Ariadne, I love you." And he signs: "Dionysus." To escape crucifixion, one must leave Christianity behind. Nietzsche comes back to the Greeks, and this time he identifies with Dionysus. The Nietzsche-Dionysus duo is up to the task, in his view, of counterbalancing the Wagner-God duo.

Finally, in January 1889, he expresses his delusion to Jacob Burckhardt, writing that what is unpleasant for his modesty is that he is unto himself all historical personages and that what remains or would remain is for Cosima-Ariadne, with whom, he declares, he practices magic from time to time. He adds that he has had Caiaphas bound in chains and that he himself was crucified the year before, lengthily and slowly, by German doctors.

I think that in Nietzsche's case we see how, starting from an absolute model, by identifying completely with him to the point of becoming delusional at nodal point N', which for the sick person consists in saying that he is even more the model than his model, Nietzsche ends up being "led" logically to say that he is God, that from now on the world revolves around him, but that, as happened to God, the crowd is turning against him, is victimizing and crucifying him. In other words, the delusion here seems to take its origin at nodal point N' and then dresses up the first brain function with perfectly logical justifications, or at least ones that are coherent within the history of humanity.

Was Nietzsche schizophrenic? I would say that from the point of view of mimetic psychopathology, he was, because Wagner was for him an insurmountable obstacle against which he would eventually be shattered, but Wagner was also a rival, whom Nietzsche's talent on the one hand, and on the other his delusion, enabled him to vanquish and surpass. The differential diagnosis between paranoia and schizophrenia oscillates and rests on these two aspects of the model. What one can definitively say is that Nietzsche, in the end, became quite simply demented. Someone demented is someone who has gone from being everything to being nothing. A demented person is someone who is no longer anything at all.

◆ ◆ ◆

What appears to me important in this new mimetic psychology is its coherence—as opposed to the DSM-IV, which is the height of incoherence: human beings are summed up in hundreds of clinical profiles. Let's be clear: either human beings are unique and then there are seven billion clinical profiles and not several hundred, or else there are universal mechanisms whose functioning produces the various clinical tableaux. True, European psychopathology distinguishes structures with different clinical forms, but our mimetic psychopathology makes it possible to understand the continuity of all symptoms and all structures through the dialectic of the three brain functions and the avatars of the interdividual relation. This explains how one can go clinically from neurosis to psychosis and vice versa, or from a normal state to a neurotic or even a psychotic one.

Let's sum up what we have seen:

· In hysteria, the model-rival is avoided because it is physically represented by an otherizing of a part of the body or by the otherizing of physiological function: a paralyzed leg or hysterical blindness, for example. The otherized body part represents the rival but at the same time hides him or her, and this rival is represented as the cause of the illness and thus as a persecutor. In any case, the true, real rival cannot be identified by the spectators or even by the patient. The latter totally misrecognizes his rivalry because it is manifested by the frenetic claim to the priority of his desire. Like Mother Superior Jeanne des Anges, in the celebrated case of the Loudon possession, the patient is "acted upon" and made sick by himself or in any case by a part of himself.
· In the psychotic process, on the other hand, the paranoiac affirms the priority of his desire with respect to the desire of all possible rivals, and he suspects everybody of persecuting him and seeking to challenge this priority.
· In chronic hallucinatory psychosis and schizophrenia, the so-called syndrome of "mental automatism" otherizes a thought or a sense perception. The thought is experienced as stolen or in any event as transparent to the Other, and the sense perception is also otherized in the form, for example, of voices that the subject hears very clearly and whose statements express the model's attitude as a rival or obstacle.

This otherizing of thought makes it possible to represent rivalry and otherness, and to complain of them while at the same time absolutely and frenetically misrecognizing the priority of the other's desire over one's own. It also translates the rebounding against the obstacle that reflects the schizophrenic's thought and auditory perceptions back at him as if they came from the other.

All of this shows the unity of mimetic psychopathology since it is clear that to represent the other-model, the other-rival, and the other-obstacle without recognizing it as real and without recognizing its true otherness, the neurotic and the psychotic otherize a part of themselves to make good on the claims at N and N' and to represent not only the model-rival or the model-obstacle itself but also the rapport that they have with that model.

One last remark: a few years ago, René Girard and I worked for a little while with a French psychiatrist: Henri Grivois. He later separated from us, and his major "discovery" is what he calls the *psychose naissante*, or "nascent psychosis." This phrase refers to a specific time when the patient suddenly feels in contact with all humans, a feeling that Grivois calls "concernment." The patient is speechless, for he is experiencing something so bizarre and uncommon that he cannot talk about it. Grivois insists that this short period will be followed by the elaboration of a delusional system or story.

I see similarities between our views: the patient suddenly becomes aware that his interdividual rapport, the relation between his third brain and the model, has evolved and is now stuck in a "rival" or "obstacle" rapport. He finds that this blockage will extend to all rapports and to all humanity. And he is speechless, unable to express anything, for a few hours or days until his first brain provides justification in the form of a delusional explanation and his second brain provides clothing: anxiety, distress, feelings of universal or mystic love or universal hatred, and so forth.[51]

CHAPTER 12

Mood Disorders

M ood disorders have gradually overrun the whole field of psychiatry and psychopathology during the last fifty years. It has come to the point that we now consider all patients bipolar and distinguish all sorts of varieties of bipolarity (bipolar 1, bipolar 2, bipolar 3, and so on), some being defined as a change of mood in the course of a single day, which obviously corresponds to just about everyone.

To take up the problem more serenely from the point of view of the three brains, it seems to me more serious to distinguish, among all mood disorders, between those that originate in the second brain, that is to say in the paraventricular nucleus of the hypothalamus, and those that originate in the third brain, that is to say in reaction to changes in the interdividual rapport. It is understood that when a human being is depressed or to the contrary euphoric, all three brains play a role in the clinical profile.

Endogenous and Exogenous Disorders

Mood disorders that originate in the second brain correspond in my opinion to what psychiatrists have always classically called "endogenous mood disorders," that is to say manic-depressive psychoses. These are characterized by a periodic recurrence of episodes of manic excitation and episodes of melancholic depression. These mood disorders are bipolar when there is an alternation in time of both manic *and* depressive episodes, and they are said to be "monopolar" when over the years the patient presents only manic *or* depressive episodes.

What characterizes endogenous mood disorders is that they come on in a recurrent way and sometimes with a remarkable periodicity in the year (for example, depressions coming every springtime or every autumn and manic states coming every winter or every summer), all of this without it being possible to detect a determinant external cause for each episode. Moreover, these disorders are recognized as having a strong genetic component and thus resulting from the biological constitution of the subject who suffers from them. Of course, they are manifested, in the case of depressions, by symptoms of a general slowing down of psychic processes, and in the case of manic states, by a significant acceleration of those processes. In both cases, the first and third brains change in unison with the second.

Thus do depressions generate a pessimism that alters the interdividual rapport and an affective anesthesia that ends up practically canceling out that rapport with one's closest friends and family. The first brain dresses up this dark mood with black thoughts presented as bearing witness to the patient's increased "lucidity." Henceforth the slowdown of psychic processes, pessimism, black thoughts, painful lucidity, misanthropy, guilt feelings, feelings of unworthiness, incurability, despair, and sometimes even melancholic delusion leading to suicide make up the communal fund to which the three brains contribute, although in this precise case the origin of the process is in all likelihood located in the second brain and in particular in the genetically programmed malfunction of neurotransmitters (essentially dopamine, noradrenaline, and serotonin).

When the neurotransmitters in the paraventricular nucleus (French: *noyau de l'humeur*, or "mood nucleus") act in the opposite way, the manic state is characterized by a gradual acceleration of psychic processes, and the subject finds himself feeling "hyped up" with, at first, increased productivity at work or in creative endeavors that are welcomed by the patient and those around him as a blessing. As the mood is euphoric, the patient suspects nothing and if the process remains tempered (hypomania), he can undertake bold projects and attract collaborators with his enthusiasm. He can enjoy considerable success, helped by the energy that carries him along, his ability to work being increased by his decreased need for sleep. But if the process is exacerbated, self-assurance will be transformed into megalomania, enthusiasm into a delusional certainty, absurd in its aspirations to omnipotence; euphoria will become impatience and aggression toward the people around him who, as they do not function as the same speed he does, are targets for his contempt and anger.

Thus in both cases, the storm unleashed by the biology of the second brain determines the evolution of the interdividual rapport in the third brain and doffs intellectual, economic, political, ethical and religious justifications provided by the first.

♦   ♦   ♦

To the group of mood disorders with an endogenous origin in the second brain must be added the repercussions on mood entailed by interdividual problems originating in the third brain. We qualify these disorders, in opposition to the others, "exogenous." For example, in a competition or a struggle, failure will elicit a depressive reaction in the second brain and a pessimistic cloaking by the first, whereas success will entail euphoria in the second brain and optimistic justifications in the first.

Thus professional failures, amorous desertions, sentimental breakups, mourning, whether the "normal" grief for a deceased person or that for a living person who has left us, and also natural failures or obstacles—illnesses, accidents, fractures, and so forth—all of these events entail a depressive state that is said to be "reactive" and that, on a purely clinical level, is very difficult to distinguish from an endogenous depression. Only anamnesis can make a differential diagnosis possible: a noticeable external cause or not and the presence of previous similar episodes. In the same way, a happy love affair, a honeymoon, a successful business transaction, a good grade on a test or an entrance exam elicit joyful and happy reactions from the second brain and flattering appraisals from the first about the self's qualities. Here too, these euphoric feelings can take on excessive proportions, but this is infinitely less common than with endogenous disorders.

Let us take a clinical example. Mademoiselle L., thirty-five years old, comes to see me about an alcohol problem. It's a strange kind of alcoholism:

every time she has to attend a meeting or speak in front of a group—and especially if her boss is present—she goes to the bathroom and drinks white rum on the sly. She has been in psychoanalysis for twelve years without appreciable results.[1] The story of her illness shows that she has mood swings, highs and lows of the sort often described as "manic-depressive" or "bipolar." I see at once that she has accumulated within herself an immense quantity of violence, which has been fairly well repressed, but which is manifested by various kinds of aggressive behavior, especially with regard to men. A few sessions later, she ends up confessing to me that she was raped by her father and that as a consequence her parents divorced. She was thus obliged to stay with her father and to take care of him. Her psychoanalyst had of course made much of this fulfillment of her Oedipal fantasies, and this only reinforced her guilt feelings.

This patient was able to become conscious rather quickly of the fact that since the rape she had been oscillating between two extreme attitudes: one was an enormous violence, consisting of hatred and a desire for vengeance against her father, feelings that were expressed clinically by a "manic" state—agitation, anxiety, hyperactivity, aggression; the other was an immense feeling of guilt that she experienced when she realized that this violence and hatred were directed against her own father. She then found herself in an impossible situation whose clinical expression is called "depression." She tried to drown her suffering in alcohol. As soon as she became aware of the fact that the two attitudes were connected, the patient felt better. She understood that her relationship with all men incessantly reproduced these two attitudes and that this was why she had been never been able to marry. She also understood that alcohol was used as a therapy in both cases, especially when she had to face up to her boss, who was a paternal figure for her. Mademoiselle L. made progress, but she was not healed. Her psychotherapy is ongoing. Her need for vengeance, which has not been satisfied, consumes her. There is a very good chance that, because her father has since died and vengeance is no longer possible, she is in reality incurable.

•   •   •

The differential diagnosis between endogenous and exogenous mood disorders requires a great deal of attention on the psychotherapist's part, and such attention is incompatible with the act of typing the various symptoms described by the patient into a computer, without even looking at the patient, and then waiting for the computer to spit out the DSM-V diagnosis that corresponds to the sum of all those symptoms. I think that if psychiatry goes down that road, it will lose touch with human beings and discount the importance it has given to the doctor-patient relationship for millennia.

A question that has always helped me distinguish between endogenous and reactive depressions is the following: "Is your essential goal, the one that is motivating your desire to commit suicide, to leave your life or to enter into death?"

Recently, a patient gave me a strange look after this question and stated that he didn't understand what I was getting at. I told him to imagine that he was locked in my office and couldn't stand the presence of the doctors around him any longer: would he be more tempted to leave the room by any means possible, the window included, or would he simply be attracted by the pavement below while looking out the window and wishing to be squashed several floors down? He answered haughtily that the idea that the pavement might attract him was the product of my sick mind and that indeed he could no longer stand the presence of the doctors around him, and that he was therefore going to ask us to leave.

My diagnosis, which was corroborated by the impression of the young psychiatrists who were accompanying me, was that not only was his depression reactive, but that his mood was on a manic upswing, given the self-assurance and aggression he displayed.

I have often been asked about the proportion of endogenous mood disorders with a metabolic, biological, or genetic origin, and the proportion of mood disorders caused by rivalry with the other or by a model who stands as an obstacle to the very desire he or she suggested to us. Answering this question is impossible, simply because psychiatrists don't analyze these processes in the same way, and thus any statistics gathering the data provided by various psychiatrists all over the world would not be homogenous and would therefore be unusable.

The chemical, pharmacological, and toxic endogenous factors capable of triggering mood disorders should not be forgotten. Thus alcohol, for example, independently of the well-known fact that some people are "happy drunks" and others "sad drunks," exacerbates mood swings, that is to say it pushes depressives "down" and manics "up." In this sense, alcohol plays an antagonistic role, going in a direction exactly opposite the one taken by lithium and "normothymic" drugs (or mood regulators), which to the contrary tend to limit wild swings in both directions. In the same way, cocaine produces euphoria and energy but, in the long run, entails states of manic excitation and delusional, paranoiac ideas. Breaking a cocaine habit, on the other hand, leads to depressive states. Hashish was used by the sect of the Assassins, led by "the Old Man of the Mountains," to trigger violence, while during the Summer of Love in San Francisco it was the drug of peace and harmony. One must therefore never neglect to verify what the patient has absorbed, medications included, since some have a depressive effect and others a stimulating one (cortisone, for example).

Addressing Mood Disorders
with Mimetic Psychopathology

What the new conception of psychiatry I am proposing aims for is to find out where, at the level of cerebral location, the problem is rooted. The most striking example is that of mood disorders, for they are perfectly localized in the second brain, in a specific nucleus of the hypothalamus. Their chemical mechanisms are largely known, and the neurotransmitters responsible for these disorders have been identified.

The causes of these mood disorders can be, as we have seen, either genetic or congenital, giving rise to a monopolar or bipolar manic-depressive psychosis, or else related to traumatizing events like the death of a loved one, a separation, or more generally the loss of an object. In endogenous disorders, it seems that the tsunami starts in the second brain and triggers a rationalization in the first ("I am guilty, I must be punished, life is no longer worth living, I must put an end to this suffering," and so forth) and a modification in one's vision of the other on the level of the third brain. This other is experienced as threatening or hostile, and then these feelings are gradually anesthetized, leading to a gradual extinction of the interdividual rapport and to the disappearance of desire. The new psychiatry will thus have to attack the problem at its root through the appropriate means, which in this case are essentially pharmacological.

It may be, however, that it is the first brain that, little by little, secretes an intellectual claim, in terms of prestige, for example, leading to an opposition to the ideas of another thinker, for example, whereby the latter becomes a rival and the interdividual rapport is modified accordingly. In this case, it is the third brain that will inherit the intellectual rivalry fantasized by the first (it was certainly Nietzsche's thinking brain that gradually had difficulties bearing Wagner's superiority).

It is certain that very often it will be difficult to determine where and how the pathological process started. Thus, for example, in hysteria, the antagonistic structure of desire is translated by an emotional and affective storm that is inscribed in the body, and it is often very delicate, when dealing with a panic attack or acute anxiety theatricalized to the extreme, to spot the mechanisms of rivalry and above all to determine if they are at the origin of the storm or are its consequences. The new psychiatry will thus have to constantly keep in mind the interaction of the three brains and try to determine on which it can act under the given circumstances and with the means at its disposal—medication or psychotherapy—and even then it will be necessary to determine the type of psychotherapy, or more exactly the quality of interdividual rapport that will be therapeutic for the patient. When I say this I think I am very close to the way of thinking of Irvin Yalom as expressed, for instance, in *The Gift of Therapy*.[2]

Endogenous mood disorders can thus trigger a tempest in the second brain, which is transformed into a tornado and then into a tsunami, sweeping away the first brain's rationality and empathy and the third brain's relationship to the other—exploding the harmony among the three, in sum.

A clinical case is a good illustration of this. I was treating a patient afflicted with a very grave manic-depressive psychosis. He was a literature professor, had a doctorate, and was remarkably intelligent. The first time I met him, he was in a psychiatric clinic, where he had been admitted after two years in an orthopedic ward: following a terrible car accident he suffered multiple fractures that required numerous surgical procedures. Miraculously, the cranial trauma that he had suffered had not resulted in cognitive impairment, such that his intelligence and memory were intact. The accident had occurred on a two-lane highway. He was driving at 130 miles per hour and had literally crashed head-on at full speed into a big rig. The psychiatric diagnosis in his file was without ambiguity: "Extremely serious suicide attempt during a probable melancholic episode."

I had numerous conversations with him on the most varied subjects: politics, literature, philosophy, and so on. A friendship developed between us, as well as a relationship of trust. One day, I said to myself that the conditions were right to bring up his accident.

"So you were very depressed. Why?"

"Your colleagues didn't ask that question. They considered that the facts spoke for themselves."

"Yes, but should I take it that you don't share their opinion?"

"Indeed. In reality, for several weeks I had been invaded by a feeling of euphoria, happiness, omnipotence. I was able to think more quickly and better than ever. I wrote with complete facility, above all during the night, because I no longer slept. Little by little, I was persuaded of my omnipotence: I thought that by driving into a big rig at 130 miles per hour I would survive simply in virtue of my quasi-divine power."

This completely upended the diagnosis. It had not been a depressive suicide attempt but rather a hyperacute delusional manic episode causing a delusion of quasi-divine omnipotence that had short-circuited reason, the emotional and sentimental part of the second brain, and erased or canceled out the third brain's empathy, the patient's concern for the other: he had given no thought to parents, spouse, or his son and thought that he would pass through the big rig as if in a science fiction film. I did my best to explain the phenomenon to which he had fallen victim, and he understood perfectly. I explained to him the absolute necessity of a normothymotic treatment, which he accepted.

I treated him for twenty years by means of this approach, completing it now with antidepressives, now with neuroleptics according to the fluctuations of his mood. One day, alas, he relapsed into a delusional mania, and the tsunami of his mood swept away everything in its path: he thought that he was stronger than everyone and didn't need any treatment. The stoppage caused his euphoria and excitation to flare up. He engaged in fraudulent enterprises and got himself cleaned out by con artists. He then stole prescriptions from my office and created false certificates that enabled him to rent an apartment. It was not long before I was called in by the police commissioner and I was of course obliged to say that the certificates were false. On the other hand, I refused to press charges and filled out a real certificate to say that the patient was gravely ill, his acts being solely the result of his pathology. He was thus released and promised me to resume his treatment. But he didn't do so and came to see me with a counterfeit copy of the *Official Gazette* (an excellent forgery, I might add) in which he was named "special envoy of the French president." I told him solemnly and amicably that he must at all costs resume his treatment, and he replied to my dismay that he was no longer in need of it, and that the president would telephone me soon to confirm his nomination and convince me that he was telling the truth.

He disappeared and for several months I had no word of him. One day, the police commissioner phoned me and announced that this gentleman had hanged himself at his home just as officers were preparing to arrest him for a serious fraud that he had committed by pretending to be acting on the French president's orders. I was terribly saddened by this news.

To the very end, he had been the victim of his delusional feeling of omnipotence, of his manic state and, seeing himself brought back down to earth, hadn't wanted to acknowledge his nondivinity and preferred to die. Was it an "anomic suicide," as Durkheim might have said? A sudden mood swing leading to deep suicidal melancholy? Or a last assertion of his absolute superiority with respect to life and death?

CHAPTER 13

Diseases of Desire

Besides the major psychopathological syndromes that we have just reviewed, there exist behavioral disorders, certain among them known in French psychiatry as "perversions," which are characterized in my opinion by the fact that desire takes on a certain independence and preeminence with respect to need and instinct. Desire, which until now was joined to need to the point that some might confuse them, takes on almost total independence here, need and instinct following along behind, incapable of resistance to such a degree that desire subverts, perverts, inverts, or even cancels them out.

These diseases of desire appear to me to stem almost entirely from the third brain, which forces the two others to, so to speak, dress it up and coif it. Everything occurs and is played out through the interdividual rapport, which is to say through the mimetic brain, because these are specific pathologies of one's relation to the model. The first and second brain functions have so little say in the matter that one can speak of neither neurotic nor psychotic "structure."

Sadomasochism

Where the hysteric plays a sort of one-man band, the sadist needs a masoch-ist to exist, and vice versa. Sadomasochism is a two-player game in which the masochist takes the role of the persecuted one and the sadist the role of the persecutor. The masochist, by his submission to the model who stands in the way as an obstacle, imagines that he (or she) is drawing closer to a surplus of being or even of divinity that he assumes is behind the obstacle. And the sadist, in the role of persecutor, finds in the masochist's submission the confirmation of the priority and superiority of his desire over the other's.

Sadomasochism is very often sexual and theatrical, but it can be purely intellectual and relational.

Sexual relations are indissociable from power relations, that is to say from "seesaw" relations. The masculine subject is able to affirm his power through erection and penetration, and the feminine subject affirms her power through absorption of the other and by the fact that the man sur-renders in ejaculation. The precocious ejaculator is one who throws down his weapons before the fight can begin. In a certain sense, he refuses the combat by refusing to assume responsibility for his role. He overvalues his partner by crouching at her feet as soon as possible.

The masochist goes still further. He submits and finds pleasure in sub-mission because submission is equated with total surrender to his partner and thus, in a certain sense, to the model, whose violence is evidence of superior-ity. As for the sadist, he finds his pleasure in the affirmation of his violence, in the fact of imposing his will on the other, and the sexual or physical suffering that he inflicts excites him because it is the proof of his superiority and of his victory, as well as of the ownership of his desire, and its anteriority, its priority over the partner's.

We can see this sadomasochistic structure of desire outside of the purely sexual arena. For example, advertising no longer has the objects it wants to promote—that is to say, the ones it suggests to the consumer's desire—pre-sented by affable, fleshy, and obviously welcoming women but by anorexic, churlish women who, as they give us the evil eye, suggest that we desire the object in question: it's as if they were forbidding us from possessing it, as if they wanted to keep it for themselves and were giving us nasty looks as they detected the glint of desire in our eyes. Advertising has evolved toward a sadomasochistic relationship with the consumer, today the model-obstacle is more attractive than the model as such, for what is forbidden is more attractive than what is permitted. And we come back to the old Genesis story . . .

What is interesting is that the very structure of the interdividual rapport, that is to say the way the third brain perceives the other as either a model, rival, or obstacle, must correspond to an individual's personal and subjective tendencies. Thus do mirror neurons powerfully reflect sexual desire when the subject watches a "normal" pornographic film, whereas certain individuals, but not all, take pleasure and react mimetically to sadomasochistic films.

Submission is even more absolute in sodomy, and that is why René Girard and I spoke of homosexuality in terms of dominant-submissive—and thus sadomasochistic, in the largest sense—power relations taking charge of sexual desire: "Desire . . . decides that the only objects worthy of being desired are those that do not allow themselves to be possessed; the only people who are qualified to guide us in the choice of our desires are the rivals who prove invincible and the enemies who cannot be disposed of. . . . After changing its models into obstacles, mimetic desire in effect changes obstacles into models. . . . Henceforth desire always hastens to wound itself on the sharpest of reefs and the most redoubtable of defenses. How can observers possibly not believe in the existence of something that they call *masochism*? . . . Desire will increasingly interpret the humiliation that it is made to suffer and the disdain that it is made to undergo in terms of the absolute superiority of the model."[1]

Girard adds: "Everyone recognizes the highly theatrical character of the type of eroticism known as masochistic. . . . The 'masochistic' subject wants to reproduce the relationship of inferiority, contempt and persecution that he believes he has—or he really does have—with his mimetic model. . . . Far from aspiring toward suffering and subjection, this imitator in fact aspires to the virtually divine sovereignty that the cruelty of the model suggests to be near at hand . . . sexual pleasure detaches itself from the object, either partially or completely, in order to fix upon the real or imagined insults that the model and rival inflicts."[2] This echoes what I said about diseases of desire, namely the increasingly important gap between desire on the one hand and instinct and need on the other. Let us note, too, that in the sadomasochistic rapport, the model seen by the masochist is sometimes a rival and sometimes an obstacle. Seen by the sadist, the image that the masochistic partner presents is always that of submissive rivalry.

I don't have many clinical examples of sadism and masochism to offer because these people do not complain of their activities, which they consider to be a private idiosyncrasy. Nevertheless, I once did have the opportunity to visit a couple who showed me a whole panoply of leather outfits, boots, handcuffs, metal belts, and whips. The woman explained to me that she was obliged to "disguise herself," as she put it, and to let herself be whipped, but not in a painful way, by her husband, who couldn't "function" otherwise. Sometime later, the husband's mimetic state had obviously become more "aggravated": the theatrical sessions had to be completed by the participation of another man to whom the husband demanded that his wife make love in order to trigger his own desire, which is of course the pure and simple reconstitution of the mimetic triangle, which Freud would obviously have interpreted as the reconstitution of the primal scene. The lesson to be learned is the reality of the dual relationship, because the woman in question ended up getting attached to one of the men whom her husband had introduced to her and ran off with him. The origin of her desire for this man was of course the husband's desire, which she had made her own. She had absorbed his desire so well that she decided to drop the model altogether and to leave with the desired object.

In another similar case, a young male patient related to me that he had been accosted by a couple made up of an old military man and a very beautiful young woman. The military man led him to their apartment, where he asked his wife to get undressed and the young man to make love to her. When a few minutes had passed, the old man shouted: "You don't know what you're doing! Get out of there, so I can show you how it's done!"

And he gave an instructive military demonstration on the art of attacking and capturing an enemy position.

Messalinism and Don Juanism

An old story has it that a gardener trained a dog to watch over his garden and guard his cabbages. The gardener was old and one day he died. The dog, loyal to his former master, continued to guard the cabbages and barked when someone wanted to venture into the garden. And yet the dog could not eat the cabbages and was uninterested in them. The prodigious Spanish writer

Lope de Vega, author of more than two thousand plays, used this fable as the basis for one of his most well-known works, *El perro del hortelano* (*The Gardener's Dog*), in which an aristocratic lady, although in love with her young secretary, refuses to marry him while also preventing him from getting married to someone else. And Molière writes in *The Princess of Elid*: "But Madame, if he loves you, you don't want anything to do with him, and yet you don't want him to belong to anyone else." This is exactly how the gardener's dog behaves.

One of the most perverse forms of mimetic desire consists in depriving the model of an object that one is unable to procure for oneself. It is in Messalinism and Don Juanism that we observe behavior of this sort. The object is in the foreground, the male or female rivals are physically absent, but their presence or the threat of their future appearance obsesses Don Juan and Messalina.[3] During the war, French peasants, when they had finished a bottle of wine, tossed it aside and said: "There's another one the Germans won't drink!" It is exactly this kind of reasoning that Don Juan and Messalina adopt when they literally empty their conquest of its substance and abandon it in a state of vacuity such that it can no longer satisfy any rival. This is the case with Valmont in *Dangerous Liaisons* by the French novelist Laclos: when he finishes with Madame de Tourvel, she will no longer be of interest to anyone. And besides, at the end of the novel she dies, and will indeed belong to no one else. In the same way, modern Messalinas, having played hot-and-cold with their lover and tossed him back and forth between hell and heaven, abandon him in such a condition that he feels like a total loser, castrated, useless, and utterly incapable of falling in love with another woman. In Don Juanism and Messalinism there is the desire to deprive not merely a particular rival but all future rivals of the pleasure to be had from an object by destroying it once it has been "wrung out." Here the rival is not singular but plural, and rivalry ubiquitous and prospective.

If Valmont is more toxic and more radical than the Don Juan of Molière's eponymous play, this is because he is not only guided by his own mimetic rivalry vis-à-vis potential rivals but also by his desire to live up to the standard of Madame de Mertueil, his mimetic model. This example is important because it shows how antagonistic desire can fuse with and be reinforced by imitation of the model as model when the two imitative desires go in the same direction. Don Juan, for his part, is acted upon only by rivalrous desire, as Molière shows us in act 1, scene 2 of *Don Juan*, in which he is jealous of the happiness enjoyed by a peasant couple he encounters: "Never had I seen two people so enchanted by each other, so radiantly in love. Their open tenderness and mutual delight moved me deeply; it pierced me to the heart, *and aroused in me a love that was rooted in jealousy*."[4] Don Juan refers to the female peasant as a "young beauty" and calls her "an utterly delectable creature." And yet it is clear that she is not so much intrinsically attractive as she is rendered irresistible by the mechanism of the model apprehended as a rival, which leads Don Juan to seduce the young woman in order to take her away from her fiancé. Here Don Juan has a visible rival, but in general, as I have already noted, the modern Don Juan seeks to exhaust the object so that it may no longer be enjoyed by any rival present or future.

Naturally the risk run by both Don Juan and Messalina is to happen upon an object (a man or a woman) that they cannot seduce. The danger is that desire would fixate frenetically on this forbidden object, on the assumption that it is forbidden by a formidable model, and that this model is in reality an insurmountable obstacle and thus alone worthy of designating the most desirable (because most inaccessible) object. In my clinical experience, however, this is not how things play out. The Don Juans and Messalinas I have encountered, when faced with the object's resistance to their charms, gave up pursuing it while simultaneously denigrating it, like the fox in Aesop's fable: "Those grapes are probably sour anyway."

All of this shows that there is no absolute fatality of desire: desire can occupy all the squares of the table in chapter 9 and have with regard to a given situation a "normal," "neurotic," "psychotic," or "perverse" attitude, but a tiny detail can make the individual jump from one square to another.

In Don Juanism and Messalinism, the second brain seems to react by providing an affective and amorous cloaking for the rival-desire's destructive undertaking. This is how Messalina seems capable of falling in love with the first person who comes along. It is not that she is actually in love, but rather that she cannot seduce if she is deprived of all feeling and really fails to manifest the least affection for the object of her desire. To the contrary, she must make her conquest fall in love with her and bring his feelings to a boiling point so that, when her first brain indicates that he has been "exhausted," she can summon the arguments provided by her rational brain to explain to the unfortunate and besotted lover the very logical reasons why she is leaving him.

It is undeniable that in the case of Messalina, abandoning the affective costume provided by the second brain and donning the first brain's implacably logical helmet contributes to the lover's destruction. The latter, in addition to feelings of abandonment, frustration, and regret, experiences a cruel lack of understanding, for nothing forewarned him of this outcome, and he refuses to accept the explanations given by Messalina, which although apparently logical and rational appear absurd to him. In his mind, nothing has changed from the night before, when she loved him passionately, to the present, when she looks on him as if he were a fly in her bowl of soup.

The same is true for Don Juan, who, from one minute to the next, abandons his unfortunate mistress, justifying himself with an "It's not my fault" that shatters what little reason the poor abandoned woman still has. In *Dangerous Liaisons*, Madame de Tourvel goes mad before she dies. All of this can be summed up in the words of the ingenious Sacha Guitry: "I'll love you for the rest of my life, tonight." In other words, tomorrow it will no longer be the case, but the one who is in love doesn't hear the last word and, as for all human beings, promises are binding only for those who believe them.

A Messalina told me that in what she called the "period of seduction" or the "start-up phase," one had to go to great lengths, be ready to make any sacrifice. In one case when she was seducing a young man, she broke her ankle, but to "hook" him, as she put it, she nevertheless insisted on making love with him without waiting for the surgical operation the following day to deal with the fracture. During this "session," she told me, she was in terrible agony, but she put up a brave front because she didn't want to mishandle the "capture." Paradoxically, this same Messalina told me that for another of her conquests, while she was in the midst of the seduction phase and thus feeling wild sentiments of love for her quarry, she nearly lost control when she suspected that her lover was continuing to have relations with his former girlfriend. She was submerged by terrible feelings of jealousy and was tempted to throw away the precious ring that he had given her as a gift. Taking on a defiant air, she told me that faced with this feeling, by which she herself was surprised, she went to bed with another "object" so as to "clear her mind" and "relativize" in all urgency the impudent lout who had nearly made her lose control of the situation by making her fall in love for real!

The comparison this story brings to mind is that of a fisherman who, pulling a wriggling fish from the water and getting whacked by its tail, takes umbrage and smashes its head with the blow of a hammer. I think one must distinguish between the rival-desire that turns against the real or imagined, current or future, actual or fictional, model, and the quivering of the fishlike lover that can annoy Messalina and cause her to brutalize him so that he surrenders once and for all.

In their day, the original Don Juan and Messalina were aristocrats. With the same techniques, manipulating others by means of the same mimetic mechanisms, and being themselves the eternal plaything of their rival-desire, they have been democratized. Their modern, vulgar avatars are the perverse narcissists examined by Marie-France Hirigoyen in her books and notably in *Abus de faiblesse et autres manipulations*, which speaks of the manipulator's "seizure of power" over his or her victim.[5] She defines the strategy of the manipulative perverse narcissist in a way that recalls the techniques of Don Juan or Messalina: "To manipulate someone, one must first seduce him, establish with him a current of sympathy and place the relationship on an 'intimate' footing based on a feeling of trust."[6] The victims of perverse narcissists are attracted by the fact of "experiencing moments of intensity";[7] this intensity "awakens" them and enables them to leave the monotony and boredom of daily life behind, even if they pay a heavy price in physical and mental health.

Impotence, Premature Ejaculation, and Frigidity

In *The Puppet of Desire*, I underlined the resemblance between two extremes of the interdividual rapport: a fight to the death, and orgasm. In other words, sexual desire, whether one likes it or not, is conditioned by relations of aggression and rivalry. In *Contributions to the Psychology of Erotic Life*, in a text entitled "Concerning the Most Universal Debasement in the Erotic Life," Freud says the same thing and characterizes psychic impotence by the dissociation of the two components of desire: "We have reduced psychical impotence to the non-convergence of the affectionate and the sensual currents in erotic life."[8] In terms of mimetic psychology, this means that desire is always impregnated with rivalry, rivalry being in some sense the "heads" side of the desiring coin, while the "tails" side is the model. The latter corresponds to what Freud calls the "affectionate current" and the other to what

Freud calls the "sensual current." To nourish sensuality and passion, rivalrous combustion is necessary, whether it is a matter of rivalry with a third party or with the other member of the couple, the first being represented by jealousy with regard to a real or imagined rival, and the second being the violent escalation entailed by the struggle for superiority, that is to say for power. When the rivals are lovers, the struggle is resolved through coitus, whose aggressive tenor is thus made evident, and the physical explosion of orgasm evacuates the aggressivity and sedates the rivalry.

Speaking of men "who are described as psychoanaesthetics, in whom the action itself never fails, but is performed without any particular pleasure," Freud adds: "From psychoanaesthetic men we are led by an easily justifiable analogy to the very large number of frigid women," and he concludes: "The affectionate and the sensual currents converge as they should only in a very small minority of civilized people."[9] From my perspective, this means that the modern world and Western culture, for lack of initiatory processes capable of taming violence and familiarizing us with death, always put us in the position of having either an excess of rivalry—that is to say, for Freud, a "sensual current" gone wild so as to make love impossible—or an incapacity to manage a normal and necessary level of rivalry or aggression, which makes the "affectionate current" so dominant that it leads to impotence.

Freud ultimately concludes, in interesting and surprising fashion, that "where they [people afflicted by psychical impotence] love they do not desire, and where they desire they cannot love."[10] This sentence is cited by Neuburger in the preface to the French edition, where he adds: "For man to fulfill his sexual urges, Freud concludes, he must necessarily debase his partner. But when he speaks of debased women, Freud cites the unfaithful woman, the kept woman or the prostitute."[11] In other words, it is always the presence of the real or supposed, fantasized, imagined rival, the one created by a delusion (as with Othello), the one suspected or simply possible, that stimulates desire and the desire for appropriation, the latter desire being one and the same with rivalry, in other words with dispossessing the rival of the object he has pointed out to us.

Well before Freud, the eighteenth-century French novelist Crébillon fils, in *The Sopha* (1742), writes that "the more tender a lover is, the less he is able to express this tenderness to those he loves. . . . The more love he has, the less he possesses the usage of the senses."[12] This passage, which concerns male impotence (as several of Crébillon's novels do), seems to anticipate Freud's notion of the "affectionate current." One could use another metaphor that I have heard from some patients: the excess of rivalry can lead to what Girard, with Clausewitz, calls an "escalation to extremes," with fury, violence, and the use of *a knife's blade* by one of the partners to penetrate the other. Or else coitus can be experienced as a penetration by *the man's dagger*, which certain women complain about, presenting themselves as the victims of the man's brutality and describing themselves, as did one of my patients, as "run through by *my husband's triumphal lance*." It is quite certain that in a case of this kind, if the woman experiences coitus in this manner, it is out of the question that she should give her lover the gift of an orgasm (her own, that is), unless she is fundamentally masochistic.

As opposed to this uncontrolled escalation, one also observes situations of total or partial inhibition, partial inhibition being characterized by premature ejaculation, which consists in surrendering to the "enemy" and throwing down one's arms from the very beginning of the combat, and total inhibition consisting in being so submerged by the "affectionate current" that respect for the woman becomes absolute and the sexual act a violation that renders the pure idol impure, dirties and debases her. The woman is sacralized to the point that she can no longer be touched.

Unlike premature ejaculation, female frigidity consists in never throwing down one's arms. Female orgasm has always appeared to me as a gift that the woman gives to the man she loves, that is to say to the man with whom she has decided to give up rivalry, at least temporarily; this is verified when lovers make up in the bedroom after a quarrel. Other kinds of women, to the contrary, can achieve orgasm at the drop of a hat, a facility that in certain cases seems to me to betray a profound contempt for their partner, who is considered as a mere accessory or a sex toy.

In his monumental work, *Frigidity in Woman*, Wilhelm Stekel prefigures the mimetic approach in a sense. He writes: "We will never manage to understand the problem of the frigid woman if we do not take into account the permanent struggle that exists between the sexes."[13] For him, the problem of frigidity is framed from the outset in terms of the antagonistic interdividual rapport between a man and a woman. For my part, I would say that frigidity is a cold war.

Some women are in a total cold war against all men and against sexuality in general. They do not even want to hear about it and an iron curtain separates them from it in an absolute way: for them, frigidity isn't a problem but a solution. For the women who come to me for a consultation, on the other hand, frigidity is a problem. For some, the blockage seems to be definitive. Wilhelm Stekel writes of a woman who had only had an orgasm twice in her life, and never since, as if she herself had decided on "anesthesia" because "veritable love would have made her submissive to the man."[14] In other cases, the cold war was accompanied by episodic or local conflicts. The woman paid the price for her efforts to satisfy her husband or lover in dyspareunia (vagino-perineal pain), vaginismus (cramps), irritations, perineal inflammations, burning sensations, urinary tract infections, and so on.

That said, Stekel observes: "In many women, the anesthesia is only apparent. In reality, *there are no frigid women*. Men's lack of experience and skill plays a big role in female frigidity."[15] In these women, the interdividual rapport is in fact "technically destroyed" by the partner's inexperience or incapacity, and the "miraculous" encounter with another partner could change things completely.

♦   ♦   ♦

One day I was called by a urologist colleague who had hospitalized a patient for unbearable pain, burning sensations, and itching in the vulvo-perineal region. He could not find any organic explanation for the symptoms and asked me to see if I could do something. I found a woman of thirty-five, tearful, withdrawn, unhappy, and suffering from pain and burning sensations that in her words were "ruining her life." Her husband was a nice fellow who didn't understand what was happening and who confessed that he had decided to abstain from any kind of sexual intercourse with his wife, since it made her suffer. When this unfortunate fellow couldn't hold back any longer, his wife surrendered "out of kindness" but derived no pleasure whatsoever herself. In this couple, the cold war was thus interrupted on occasion by "hotter" conflicts now and then. I tried to explain to the patient the psychogenesis of her issues, but she didn't want to hear me out and, like all neurotics, insisting on the reality of her physical symptoms and her suffering, even suggested that I should consult with her urologist if I didn't believe her. I realized that her rational brain didn't want to see anything and that her mimetic brain wasn't ready to recognize the conflict. I then had the idea of making her laugh with a war metaphor inspired by that morning's newspaper: Margaret Thatcher had just sent the English fleet to take the Falkland Islands back from Argentina. Going into the patient's room, I asked her, "So, how is the situation in the Falkland Islands this morning?"

"What are you talking about?"

"You told me that you had experienced a flare-up in this region. I decided to call it the 'Falkland Islands' because war is raging there, and yet it's a region where nobody wants to go!"

She burst out laughing.

"What strange ideas you have!"

A barrier had fallen, and we were able to speak much more freely about her sexual problems. It became possible for me to suggest some tranquilizers, to calm the emotive and eruptive reaction of her emotional brain, without the risk of resistance on her part. The pain and various other symptoms abated in a few days and she agreed to leave the hospital, on condition that I prescribe to her husband total abstinence from now on. The poor guy agreed and the patient disappeared for a time.

Three months later, she came to see me and said that her husband, who couldn't take it any longer, had asked for a divorce. She expressed surprise: "It's strange, I love him and yet I feel relieved!"

I made no commentary and encouraged her to concentrate on her work, without stopping her medication. Six months later, I saw a completely transformed young woman walk into my office: radiant, with a new hairstyle and makeup. She was chic and even "sexy"—I hardly recognized her.

"Guess what happened!" she said. "I met a man, a real man, and with him everything changed. We never stop making love, and I have orgasms, I'm happy. I'd never experienced that before."

"Well done! And I assume peace has returned to the Falkland Islands?"

She burst out laughing: "Not only that, but it's become a delightful vacation spot!"

◆　◆　◆

In other cases, the mimetic cold war takes the form of nymphomania. Stekel reminds us that as no man and no quantity of men could satisfy Messalina (Claudius's aforementioned third wife), she was nicknamed "Invicta" ("unvanquished"). With this type of woman, Stekel tells us, "The desire to dominate becomes much more important than the joy of orgasm. She values her personality more highly than the intensity of the sexual relationship."[16]

One of my patients boasted of being able to make all men experience intense pleasure, of submitting them to her desires, without ever feeling anything herself except the pleasure of conquest, of victory, of the submission of the men who fell in love with her. One day, she came to my office in dismay.

"It's catastrophic," she said. "I have fallen in love, and I've fallen hard."

"Bravo," I said. "You must be happy."

"Are you an idiot or what? This man discovered how to give me orgasms. Believe it or not I have vaginal orgasms. I would never have guessed it, and I'm furious!"

"But why?"

"I don't want to become dependent on him, or to be indebted to him. So I left him immediately to sleep with several guys with whom I have orgasms, but only by applying a technique that I've learned, and while directing them and using them."

Frigidity and impotence both are diseases of desire, a frenetically rivalrous and obstinately obstacle-driven desire.

Anorexia

Anorexia and bulimia (which I'll discuss next) are diseases of desire. That is to say that in these pathologies, need, in exemplary fashion, is totally dissociated from desire, in the sense that a need as fundamental and basic as nourishment is completely evacuated when competitive passions take precedence over physical and psychological health.

Mental anorexia has been defined as a nosological entity characterized by three phenomena: anorexia, emaciation, and amenorrhea. It usually afflicts young women, but young men are increasingly affected, too. It is clear that the young anorexic's ferocious desire is able to override need, instinct, and physiological balance to the point that the three aforementioned phenomena are exclusively somatic.

In what sense is anorexia mimetic? First of all in that it comes from a desire to imitate the models of feminine beauty that our ambient culture and advertising offer up to us. As I have written elsewhere, fashion "models" obviously deserve their name: "Mimetic analysis suggests that the current epidemic of anorexia is a contagion, propagating a filiform standard of beauty among adolescent girls and causing them to become obsessed with the need to lose weight in order to look like the ultra-slender goddesses of the cinema, television, and glossy magazines."[17]

In his book on anorexia, Girard offers an additional insight into the problem when he underlines the importance of competition, that is to say of rivalry. Girls who deliberately fast are seeking to resemble supermodels, it is true, but this is not because they want to attract men.[18] No man finds a skeletal woman attractive, but women don't care: they are seeking not to conquer a handsome man but to outdo their feminine rivals in skinniness. Girard writes: "Anorexic women are not interested in men at all; not unlike these men, they compete among themselves, for the sake of competition itself."[19] It is in this sense that anorexia illustrates in dramatic fashion the strength of rival-desire. Skinniness becomes an end in itself, detached from all concrete benefit.

The anorexic gets locked into the escalation of mimetic rivalry by striving to become ever thinner, and thinner still, by going further than all her models. In this sort of "fight to the death," the object, once again, disappears, has no more importance, is lost to sight, annulled: hunger, instinct, need, nourishment, sexuality, and men no longer exist. Food items become horrifying because they are so many tempting obstacles strewn in the path of victory over the model. They must thus be eliminated.

After the second and third brains, the soma, the body itself, is drawn into the battle. Body image becomes quasi-delusional: the emaciated patient asserts that she is fat! The body is lost to sight, sacrificed on the altar of rivalry: amenorrhea, emaciation, and in the long run the disappearance of secondary sexual characteristics ensue. Parents, friends, and doctors who cry and beg are violently swept aside and manhandled, treated as enemies, for their love, attention, pleading, and tears are experienced as so many obstacles to the final victory of rivalrous desire.

I think that the anorexic also seeks to take power over the members of her family. The latter become prisoners, obliged to watch over the patient and to count along with her the calories she consumes . . . or doesn't. The young anorexic woman thus becomes the center of familial attention, and as I have written elsewhere, "Her plate becomes a sort of Roman circus where the rival-desires of those around her—who want her to eat—clash with her own desire, her refusal to do so, which has the whole family on tenterhooks and keeps it involved in the daily struggle, until recourse to 'medical authority' signals the defeat and resignation of her parents and the appearance, at last, of a worthy adversary."[20] Alas, most of the time this adversary, too, will be defeated.

Bulimia

Bulimics, from my point of view, are of two sorts. Some are anorexics in that they want to become thin, but their rival-desire is too weak, is not absolute, is in favor of compromise, and wants to have its cake and *not* eat it too: they "chow down" and then make themselves vomit. They feel guilty, are unhappy, lost, depressed. In these patients, the second and third brains resist and a grave disharmony comes to light among all three brains. This internal struggle exhausts them, and the second, emotional brain gets depressed. The doctor and the psychotherapist have a chance to help them if they can reinforce the first brain's arguments and first calm and then reorient the energies and moods of the second. Finally, they must attempt to defuse the third brain's rivalry.

The other variety are obese bulimics, who eat to fill a void and to appease their anguish. In these patients, desire exaggerates need and erases the normal instinct of satiety. Permanent hunger is an ersatz of desire, which makes them similar to drug addicts. However, the accumulation of extra pounds recalls obsessive hoarders, for whom the model is an obstacle that they keep rubbing up against. This type of bulimia and obesity seems to accompany the destructuration and decadence of certain cultures. These obese people are very numerous in the United States and are starting to multiply in Europe from childhood on. Here, the third brain, chained to a desire that is never satisfied, gets help from the first, which ritualizes permanent chewing and swallowing, any interruption in which rapidly brings on anxiety produced by the second brain.

Drug Addictions

Today's world and current culture are devoid of taboo. That is why, in my opinion, young people today are experiencing a lack of desire. Nobody has forbidden them anything in this permissive world and they have thus been deprived of the fundamental human experience—that of desire. So we shouldn't be surprised that the number of drug addictions has multiplied. What many young people seek in drugs is an ersatz of desire: lack. These youngsters are not so much after pleasure or artificial paradises as they are desperately in search of an experience capable of engendering in them a self that for the moment is nonexistent. They are seeking to live through a structuring experience of desire that the absence of taboos and authority and the laxity of their parents have prevented them from having.

Today's drug addictions are the fruit of a world without taboo, a world in which change is occurring so rapidly that a gulf is created between the generations. Adults and adolescents no longer understand each other, and yet the adults behave and talk like young people while adolescents enjoy a quasi-adult independence in their financial and sexual lives. Starting in the 1960s, young people became conscious of the fact that the life that was awaiting them would be so different from their parents' lives that to survive and find their way they would have to invent their own manner of seeing the world, apart from any sort of transmission of knowledge from parents to children. They had to invent a completely original philosophy of life and politics. And, as I wrote forty years ago in *La personne du toxicomane*, "The first originality that comes to mind is of course to do exactly the opposite of received values."[21]

This originality can be authentic, but it rarely is, and the more it seeks to deny the model constituted by the parents' generation, the less it manages to blaze a truly independent trail. The gap between the generations hides a latent similarity. The abdication of adults and the resulting absence of authority removes the one source of stability capable of guiding children through life. This absence of authority is manifested by a certain number of symptoms that Jean-Marie Sutter summed up as follows.[22]

· Weakness and inconsistency of the personality, which is deprived of a framework, lacks lines of force, and is governed by occasion and caprice, with a moral sense that seems deficient or anarchic.
· Insecurity, anxiety always just beneath the surface, with frequent recourse to suicide in the presence of even the most banal difficulties.

In light of these symptoms, nosological categories are so vague that their validity can be neither denied nor confirmed. The diagnosis of "psychic imbalance" doesn't tell us much of anything about drug addicts. To understand them, they must be likened to participants in ancient initiatory rites that have fallen into disuse in our modern society. This product of the isolation of young people today "covers up a thirst for initiation that Western culture and society can no longer quench, because *rites of passage* have disappeared from our civilization."[23] The ones that had to do with sexuality, for example, have been swept away by the sexual revolution. "Young people no longer encounter any difficulty in being initiated into sex, no obstruction, no taboo that might make restore to the discovery of sexuality its value as a rite of passage. Where sex is concerned, there is no longer any possibility for transgression."[24]

In the 1970s, when I was writing my book on drug addiction, sexuality in Spain, Italy, Greece, and Turkey was bristling with taboos that gave sex all its spice. In these countries, sexual initiation still absorbed most of the energy expended by adolescents. It consisted in surmounting taboos and obstacles that veiled the mysterious—and thus desirable—thing, and which quenched the thirst for initiation even as it exalted desire.

Things have changed a great deal since that era, and the upheavals that shook Western countries after World War II, beginning in the 1960s, have transformed mores and attitudes in prodigious fashion throughout the world, such that taboos are eroding and drug addictions are multiplying. Let me repeat myself: drug addictions must be considered as substitutes for rites of passage, insofar as they provide young people with the experience of desire in the form of *craving* and thus enable them to experience both the heights of ecstasy and the depths of physical and mental annihilation. "Since our civilization doesn't have rites of passage to offer its youth, young people are left to face their anxiety alone. For the primitive individual, this terrible experience of anguish was indispensable for the birth of a new man: no initiation was possible without a ritual agony, death, and resurrection. Young people in our day have felt this essential need rising up in them from the depths of the past."[25]

Current drug addictions are an illness due to an agenesis (absence of development) of desire. If craving is a substitute that makes it possible to have an idea of what desire can be when one has never experienced it, it is also a desperate attempt to create a self insofar as, because desire creates the self, when desire doesn't exist, the self doesn't either, whence the enormous difficulty that psychiatrists and psychologists encounter in attempting to make an accurate diagnosis, that is to say in attempting to characterize the absent or evanescent self of these drug addicts.

Desire, as I have said, is movement and energy. Its finality—that is, the choice of its object—is constituted by parental models and taboos. The latter strengthen desire, in the same way that muscular strength is developed by meeting with resistance. The laxity and abdication of our civilization, by depriving young people of taboos, is preventing them from giving their desire muscle, so to speak, so that it can be transformed into willpower. In my anthropology of the self-of-desire, this agenesis of desire entails a non-formation of a solid self, wherein atrophied desire engenders a ridiculous, feeble self. This is where drug use finds its justification.

Desire that is stubbornly exercised against obstacles and difficulties gains "muscle," so to speak; it is solidified and transformed into willpower. Willpower is desire that, once adopted, is maintained in time until it is realized, keeping up its strength all the while. And this type of desire, which has been transformed and solidified into willpower, is opposed to the fluttering, floating mimetic desire that changes objects and objectives according to circumstances and the nearest models at hand. For such a desire, no obstacle is discouraging and absolute, and between the moment when desire is solidly adopted and the moment it is fulfilled, many obstacles will be overcome, beginning with the wear and tear caused by time. For example, Nicolas Sarkozy's and Francois Hollande's desire to become president of the Republic survived years of various difficulties and obstacles and proved able to resist all the shocks strewn along its trajectory. During these difficult periods, the second brain was in great distress, but the third and the first were allied in the quest to make good on a desire that over the years had become unshakeable.

◆　　◆　　◆

We are thus witnessing the emergence of new phenomena that I have baptized diseases of desire, that is to say illnesses of relationship, in other words illnesses that are purely cultural and not natural, and that are linked to our modern, Western society. Anorexia and bulimia do not exist in poor or developing countries and only exist in the most opulent Western countries. Current drug addictions likewise reflect a new and original cultural use of drugs that are as old as the world. The delinquency caused by these drug addictions varies according to the cultural and social milieus in which it occurs.

PART 4

The Dialectic of the Rational, Emotional, and Mimetic Brains

CHAPTER 14

The Mimetic Mechanism

A s I have explained, this book does not aspire to answer the "why" question. This question is a trap into which many psychologies and psychopathologies have fallen. I understand that trying to determine once and for all whether the cause lies in the first, second, or third brains or in a malfunction common to the three is a useless and even a dangerous undertaking. Indeed, this undertaking would lead immediately to physiologists, chemists, and neuroscientists becoming involved, on the one hand, and on the other end of the scale, metaphysicians and all sorts of religions. My goal is thus clinical and phenomenological observation, that is to say, trying to answer the "how" question.

It seems to me that "why" is an immature question that children never stop asking. We must become adults by renouncing this question and being content to modestly study the how: how things happen, how phenomena fit together, how disorders are manifested, how the mimetic mechanism evolves first toward rivalry and then toward the obstacle, causing neurotic, psychotic, and perverse symptoms, or somatic upheavals, as we have seen. But also how to react and how to behave clinically when faced with human problems.

To do this, we will have to use a new form of psychotherapy that integrates all the notions we have spoken about, while at the same time not depriving ourselves of medications. We are already familiar with some medications that have the first brain as their target (neuroleptics), and others that aim at the second brain (tranquilizers, antidepressants, normothymic drugs). The challenge of pharmacological research, it seems to me, consists in finding medications whose target would be the third brain. That being said, no medication can substitute for love and authority, for the experience of what is allowed and what is forbidden that can be provided by a harmonious education that results in a desire solidified into willpower and produces a structured self.

Now let us look at some general examples illustrating the dialectic of the three brains.

A man walks into a bar and hesitates before ordering. Another man arrives. In a strong voice, he orders a pint. At once, the first man, acted upon by his mirror neurons, addresses the bartender: "Me, too."

The third brain took the other as a model and copied him. The first brain is limited to giving the choice its stamp of approval by placing the order, and the second brain produces feelings of satisfaction at the idea that the decision has been taken and that a good beer is on its way.

Taking the other as a model is positive, in this case, because it is not damaging. Nonetheless, it amounts to taking the path of least resistance, and the mimetic desire unleashed by our civilization is always tempted by this facile way out. Thus do human beings, like Panurge's sheep in the novel by Rabelais, rush after one another without wondering where the movement will take them. They follow the movement triggered by their third brain without asking any questions because it's easy. How difficult it would be to lift up one's head at the bidding of the first brain and look to see where one is going, to perceive that mimetic desire will take us over a cliff into the ocean and thus to death, to mobilize the energies, the emotions, and the strength of the second brain's feelings to part ways with the herd! But no: when it is a question of choosing between difficulty and death, the path of least resistance leads to death.

Today, the political powers have understood this. In the past, politics "from above" was the rule: the king chose his successor, just as Richelieu "passed on" Mazarin to Louis XIII. Today things are different: democracy and universal suffrage have made "from below" the key to politics. One must therefore do everything possible to keep Panurge's sheep moving and to make them believe that the path of least resistance will lead them to the immediate fulfillment of their desires. "Change happens now" (*Le changement, c'est maintenant*) is a marvelous formula in this respect, and helped ensure that current French president François Hollande would get elected. The third brain of a majority of French people was guided by a mimetically shared desire for "change," a term that is sufficiently vague to encompass all particular desires, and that implicitly contains the promise of their immediate fulfillment. Indeed, change, in this sense, is a fulfilled desire. Who hasn't heard someone say: "If Juliette said yes, it would change my life," "If I got this job, it would change my life," and so on. Milton Erickson taught that one must always use vague terms and notions in hypnotic induction. And in the political realm, Talleyrand affirmed that one always departs from ambiguity at one's own expense.

In a crowd situation, the third brain is engulfed by a mimetic torrent. It is enchanted by facility. It invites the first brain to go into sleep mode and reassures the second, which basks in euphoria. In a one-on-one situation, to the contrary, the third brain, when it takes the other as a model, produces empathy, understanding, friendship. It mobilizes the first brain for the purpose of apprenticeship, copying the model and learning what the model teaches. The second brain is charged with furnishing feelings of joy, enthusiasm, love, and gratitude, as well as a good, positive mood.

◆　◆　◆

Now let us take the example of a man (or a woman) driving a car into a parking lot in search of a space. At the moment he or she sees one, another car shoots out of nowhere and slips into the coveted spot. The third brain at once identifies this intruder that took the parking space as a rival. The second brain is furious, and produces feelings of anger and a threatening mood. It is left to the rational brain to calm things down and decide to keep driving and look for another spot.

But our reason doesn't always have the last word, as it did in the parking lot, far from it. More often, it is pulled along by the third brain and requisitioned to justify the rivalry, to accuse the enemy and make him bear the responsibility for the conflict. As for the second, emotional brain, it provides appropriate feelings and emotions and produces a depressive or expansive mood depending on whether the conflict is going well or badly.

The exacerbation of rivalry can also "escalate to extremes" and put the first brain out of commission, even as the second causes passionate feelings of jealousy, hatred, and violence to flare up. Drawn along reciprocally by each other, the third brain, which is "stuck" in the rivalry position, and the second eliminate the reasonable influence of the first brain and the escalation of rivalry ends up so absorbing the protagonists that they lose sight of the object of their rivalry, because they are literally fascinated by each other and by the rivalry that unites them. At the time of this writing, an example of this was provided by Valérie Trierweiler's tweet, which caused a media firestorm in France: Trierweiler, the partner (at the time—they have since separated!) of French president François Hollande, took to Twitter to declare her support for the electoral opponent of Hollande's former companion, Ségolène Royal, whom Hollande was publicly backing in a socialist runoff vote. Trierweiler's tweet caused the president public embarrassment and was widely ascribed to jealousy. In this case, the two rivals—Royal and Trierweiler, but especially Trierweiler in this episode—lost sight of the object of their rivalry, which was nothing less than the president of the Republic himself.

Mimetic rivalry flares up on the collective level just as it does on the individual level. Long ago, two furious women were introduced to King Solomon: shouting and inveighing against each other, they both asserted that the child that had survived was theirs, and they tried to snatch it away from each other. The wise king decreed that the child should be cut in two and that each should have one half. At once one of the women, for love of her child, renounced mimetic rivalry and the object of her desire, and abandoned it to her rival so that it would remain living. The king concluded that she was the true mother. In 2012 in the French conservative party (the UMP), two politicians, Copé and Fillon, fought over the leadership of the party in the same fashion, but as King Solomon was absent, the baby was cut in two: after losing the vote, Fillon created his own UMP faction in the French parliament.

Let me take this opportunity to specify once more the model's fundamental role in the interdividual rapport that, through the interplay of mirror neurons, constitutes what we have called the "third brain" and the evolution of the other-model toward a model-rival situation. I would like once again to call on Jacques Lacan as objective witness, since I myself have never been Lacanian. Lacan uses the word "prototype" where we use the word "model," and, speaking of the paranoiac delusion, a delusion that represents the transformation of the model into a rival in a psychotic structure, as we have seen, he writes: "We have already underlined in the analysis of the delusion the double, triple, and multiple character that the persecutors present in their role of symbolizing a *real prototype*."[1]

The "persecutors," that is to say the rivals, are replicas of a real model, a "real prototype," experienced—sometimes with good reason—as a maleficent rival. In the case of "Aimée," a patient Lacan observed and about whom he writes at length, it is the sick person's cherished sister who is at once her model and rival, and everything that recalls this prototype becomes, through an "iterative identification,"[2] a rival and a persecutor. Thus, if the prototype or model is a magistrate, all magistrates will be enemies and persecutors; if it's an actress, all actors and actresses, in the theater or on television, will become, through an "iterative identification," persecutors as well.

What is remarkable is that it is the circumstances of life that create models through the instantiation of the mimetic interdividual rapport, in virtue of the automatic interplay of mirror neurons, and it is mimetic reciprocity that then makes the model evolve into a rival or an obstacle.

◆　◆　◆

Here, for example, is a model becoming an obstacle right before our eyes. It's a joke that people used to tell during the Soviet era. The soldier Popov meets another soldier, Ivan, and the discussion begins.

"My dear Ivan, you're never going to believe what happened to me!"

"What is it, Popov? You seem all shaken up."

"And for good reason. You know our general, General Domakine, whom we hold in such high esteem."

"Yes?"

"Yesterday evening I came home earlier than planned and what do you think I saw in my bedroom?"

"What?"

"The general making love with my wife."

"The general? Such an important and respectable man? With your wife, the one you love and married six months ago? My poor Popov, what did you do?"

"I was very lucky: the general didn't see me! And I left without making any noise."

"You were lucky!"

In this situation, the general, the absolute model for the two soldiers, becomes an absolute obstacle for Popov in the blink of an eye. He stands as an insurmountable rampart between the soldier and his wife. Circumstances are such that Popov can never see him as a rival, and his first brain is insightful enough to make him see this. The third brain therefore puts itself in the "obstacle" position and gives up, while the second brain accompanies Popov's flight with ambivalent feelings of relief, chagrin, anguish, and a depressive mood. Popov doesn't even think of being jealous, since jealousy is a feeling that the second brain furnishes only when the third is in the "rival" position.

CHAPTER 15

Some Clinical Studies

I n a clinical setting, it is in fact always the second, emotional brain that wants to express its emotions, feelings, and suffering. It falls to the psychiatrist to try to determine the dialectic at work among the three brains, to detect if possible in which one there is an avenue for treatment (whether pharmacological or psychotherapeutic) to modify the pathological organization. He will observe, in the vast majority of cases, as I have tried to show in this book, that the problem originates in the third brain, that is to say in the patient's rapport with others. He will then have to determine whether the model is a rival or an obstacle, after having—if possible—identified it. He must then evaluate the reactions of first and second brains: in other words, make both a classical psychiatric diagnosis and a mimetic psychopathological diagnosis. Finally, he has to decide which of the three brains should be the target of a pharmacological treatment and what psychotherapeutic approach will best untangle the crisis.

All of this is neither easy nor simple. Let's take some examples from my clinical practice.

The Human Football

Violette is a young woman of twenty-seven, from a good background, intelligent, and she introduces herself in a direct, open manner. She hands me a letter from the doctor addressed to me, which informs me that she has just left a rehabilitation clinic where she spent three days following attempted suicide by overdose of painkillers.

After having read the letter, I look at her with astonishment. The question is expressed in my gaze and she answers it.

"I couldn't take it anymore. My parents have been involved in divorce proceedings and fighting with each other for years. Both of them try to get me on their side and accuse the other one of all kinds of things. I love my father, I admire him, and I am afraid for him because he has a weak heart and he's not as young as he used to be. But I work in my mother's business, and she is a remarkable woman. I don't understand her. I don't understand why these two great people whom I love hate each other, why they make me a witness to their brawls and their inner feelings. I was exhausted and I wanted it to stop . . . So that's why."

"I understand," I said. "Your parents, fascinated and absorbed by their rivalry, lost sight of your feelings, your interests, your affectivity. And yet both of them are attached to you and each wants to convince you and keep you close. You have thus become one of the stakes in this rivalry, perhaps the principal stake. I am going to draw a comparison. In a soccer match, the two teams are fascinated by their rivalry and what's at stake is the ball. But the players don't give any thought whatsoever to the 'well-being' of the ball: it's the ball that gets kicked around and that suffers the most over the course of the match. In the match your parents are playing against each other, you are the ball!"

"Yes, I understand all of that. But I am tired of being the ball, as you say. What referee will blow the whistle to end the game? When will the lawyers and judges finally put an end to these interminable procedures?"

"Your suicide attempt was an emotional reaction due to discouragement, disgust, and exhaustion and it was authorized by your intelligence in hopes of 'waking up' the protagonists and forcing them to become aware of the damage their conflict was doing to your mental balance. I think you have achieved your aim, but it won't suffice to 'heal' them. They will get sucked back into their rivalry."

"What should I do?" she said.

"I don't think you are depressed, in the sense that I don't think that antidepressants can help you. On the other hand, I think that a light sedative could reduce the intensity of your emotions and help you to take a step back from the situation. And above all, we are going to work together to change 'sports': we're going to try to make it so that from now on you watch your parents playing a game of tennis and follow the ball as a spectator, so that you won't be shaken and buffeted by each of their movements."

The outcome was positive and Violette even discovered that watching a tennis match was more fun with company: she found a boyfriend who helped her a lot—at last, someone was watching out for and caring about *her* feelings.

The "Peace of the Valiant"

André, an intelligent, dynamic, and successful CEO, enters my office, takes off his jacket, loosens his tie, and tells me, "I'm in a shitty situation!"

"How is that?"

"I had a liaison with my former secretary, and a few days ago my wife discovered everything by looking through the messages on my cell phone, which I'd left lying around. Ever since, it's been hell. My wife shouts at me from morning to night, even in front of the kids, reproaches me incessantly, and is threatening to ask for a divorce. I've told her over and over again that it's ancient history, that it's over, that I haven't seen Mathilde for months, but we can't spend an hour together without her harassing me with questions: where, when, how, how many times, and so forth. If I don't answer, she imagines the worst, and if I do answer, she builds on my answer and keeps digging and there's no end to it."

"Let's be clear: do you still love your wife and does she still love you, in your opinion?"

"The answer to both questions is yes—yes, *but* in reality I think I'm still in love with Mathilde. I fired her and she's been working for another company ever since, but she tells me that she loves me, that she will wait for me, that she understands me, and when we see each other, she tells me that I am a coward, that I am afraid of my wife, that I don't dare ask for a divorce. And so for months now I've been getting yelled at by the two loves of my life, neither of whom will make love with me, and I am completely at a loss."

"Let's take things in order: are you able to work, to read, to concentrate, to eat, to sleep? Are you having panic attacks, episodes of uncontrollable weeping?"

"Yes, I am anxious. I'm having a lot of trouble getting to sleep at night, all the more so because my wife dresses me down every night, and moreover, I wake up in a sweat at three or four in the morning. Yes, sometimes I start crying, all alone at the office, and my new secretary, who is very understanding, is the only witness to these attacks. She helps me a lot because I forget everything, and she has to keep reminding me about appointments, and even about my own decisions with regard to certain deals."

It is clear that this patient's third brain is oscillating continually, with regard to his spouse and Mathilde, between the "model" and the "rival" positions. He loves each of them differently but fears and detests them as intensely as he loves them: his second brain, called upon to produce contradictory feelings and changing several times a day, is in a state of disarray that is blocking the first brain, by soliciting and holding its attention constantly with a new emotion, feeling, or mood. Going back and forth from love to hatred and from hatred to love, the third brain is causing the other two to panic and is plunging them into confusion.

"Let's sum up," I tell him. "I think that the first order of business is to calm you down, to calm your nervous system so that you can reflect, react, and make decisions. So I am going to prescribe you a light treatment of tranquilizers and antidepressants that will take away your anxiety, your urge to weep, and enable you to sleep. At the same time, I suggest that you explain to Mathilde that you are in treatment and that she should neither see you nor telephone you for a while. Ask your wife not to talk to you about Mathilde anymore until you are feeling better and so that you can reflect. And invite her to reflect, too."

A few days later, this patient came back to see me and, under my interrogatory gaze, tells me: "I'm doing better. I am less anxious. I am eating and sleeping better. Mathilde reacted well and told me that she was sorry for me and that she understood, which hasn't prevented her from inundating me with text messages. As for my wife, she has accepted the idea and is exploding less often. On the other hand, she tells me that she wants to come to see you to tell you the 'truth,' in other words her point of view, for fear that you are taking my side!"

I note that the tranquilizers have reduced the level of anxiety and that the second brain has been soothed. The antidepressant has improved mood and sleep. The second brain is now allowing the first to reflect, to function, and to concentrate, not yet perfectly, but better. The third brain is still in oscillation, but it too is benefiting from the calming feedback of the second brain. I tell the patient that the treatment must be continued, that I am of course ready to speak with his spouse, but that nothing will be definitively resolved until he has made his decision and chosen clearly between the two women. He laments this and expresses regret that circumstances have not allowed him to keep them both! I answer him that I understand his desire but that this desire is unrealistic given that both women want him all to themselves, that they know each other and that they are keeping track of each other.

Sylvie, his wife, who has just entered my office, is a pretty, elegant woman who gives me an intelligent but mistrustful look.

"I think that we have a problem," I tell her, "and that it is good that we are talking."

"I want to make things clear. My husband and I have been married for twenty years. We have three children who are eighteen, fifteen, and ten. The secretary seduced my husband and he was clueless. He thinks that she loves him when in fact she is after his money. It's very hard for me to forgive him. I am scandalized by his attitude—how could he? I am thinking about divorce but I can't make up my mind."

"The only true problem is, do you still love your husband?"

"I love him and at the same time I hate him. It depends on the moment. Of course I love him, but every time I think about Mathilde, I yell at him and I get carried away and even hit him or throw a glass of water in his face."

"Your rival is Mathilde. Don't confuse adversaries. If you want to keep your husband, don't take him as a rival, don't treat him as an enemy, a guilty party, don't cover him in reproaches and insults. This attitude goes against the goal you are pursuing: you want to keep him and triumph over your rival. You must therefore get closer to him and ally yourself with him against her. But by constantly yelling at him you are going to make him run away from you and maybe to her."

"I understand what you're saying, but you are a man, you want to defend him. Do you think that he deserves my sympathy after what he did?"

"It's up to you. And I am sorry to say that the effort has to come from you. He can't do anything: he has been begging for your forgiveness, has abased himself, broken up with his mistress, and he swears that he loves you. You have to decide what victory you wish to obtain: victory over your rival or over your husband. Both have advantages and disadvantages, but for my part I think that victory over your husband would be a defeat for your marriage, a catastrophe for the children, and a personal misfortune for you. I am on nobody's side. It is impossible to go back in time and it is possible to reason 'negatively'; what solution would be least painful for you: to crush, humiliate, and leave your husband to punish him and lose him, or to keep him while eliminating your rival but on condition that you overcome your pride?"

Sylvie tells me that she is going to think about it but that she thinks that all men are the same.

"Maybe I should consult one of your female colleagues," she adds as she is leaving.

A few weeks go by before the husband comes back to see me.

"I have continued the treatment and it's helping me a lot. My wife yelled at me after leaving your office, but I have to say that since she saw you her attitude is incomprehensible: she insults me, calls me all sorts of horrible names, and tells me that she wants a divorce, and a few hours later she throws herself at me with a passion that she hasn't shown in a long time and we make wild love."

I realize that Sylvie's mimetic brain is oscillating between rivalry and love-as-passion. Earlier, I underlined how antagonistic and violent their love is. In reality, Sylvie is oscillating between two ways of expressing her rivalry: struggle and conflict on the one hand, and on the other possession and appropriation and thus in a certain sense domination. In the struggle, the second brain furnishes feelings of hatred and a dark mood, and the first moral, social, familial, and pejorative arguments against the husband. In passion and orgasm, the second brain furnishes feelings of triumph and a rosy mood, while the first confirms the victory. Here we have the workings of the interdividual rapport in a hysterical mode: the crisis mimes now a fight to the death, and now orgasm. The third brain seesaws back and forth between the two attitudes and they are manifested at different times. To prevent this from becoming chronic, there is still work to be done.

"And you," I say to André, "have you made your decision? Do you want to keep your family or are you still attracted by Mathilde?"

"Of course, I want to save my family, and moreover I love my wife. The passion that she has shown me from time to time makes me desire her more and more and that makes the moments when she's haranguing me all the more unpleasant. I don't know what's going on anymore. As for Mathilde, I am avoiding her, and I would like it if she met another man, someone who was available, so that she could be happy. Sometimes she scares me: I am afraid that she is going to guilt-trip and yell at me, I'm afraid she's going to phone my wife, and I'm afraid she's going to kill herself. I'm afraid of everything!"

"Did your wife consult with a female psychiatrist?"

"No! But she wants to see you again and asked me to make an appointment with you. I don't know why."

A few days later Sylvie arrives at my office.

"Why are you taking my husband's side?"

"I am taking nobody's side. I am trying to be on the side of reason."

"And what does that mean?"

I realize that during the last session I didn't manage to connect with the third brain. Or at least not completely, because she has a fluctuating interdividual rapport with her husband, now antagonistic, now passionate and amorous. This time I decide to address her first brain.

"You are an intelligent woman. You are winning the fight against your husband, because your husband is still there, he is in love with you, and he wants to win you back. Your problem is overcoming the secondary and derivative rivalry that you have with your husband. You want total submission, an unconditional surrender. You've almost got it, but that's not enough for you. He is in a state of inferiority, guilt-ridden. It is up to you to raise him up and make peace. I am going to take an example that may seem strange to you. When General de Gaulle wanted to make peace with Algeria, he spoke to the Fellagha, to the ex-convicts and the assassins, as if they were recognized combatants, and he offered them the "peace of the valiant." He raised them up to his level instead of despising them and condemning them. He put them on the same level as the French soldiers: valiant soldiers. This shocked those who considered them rebels, but it made peace possible. In the same way, you must respect your husband, raise him up to your level, recognize his weakness but also his efforts and start a new life with him, for you both and for your children. To do that, you must give up the vengefulness that is eating you up and set aside your pride, which has suffered. But you will have saved your family. I can place my trust only in your intelligence: it alone can save this situation."

Sylvie looks at me for a long time, gets up, and goes to the door. As she is about to leave, she turns around and tells me, "You're asking a lot."

"Yes," I say, "more is always required of the victors than of the vanquished."

I spoke essentially to her first brain and, at the last moment, I flattered her third brain to encourage her to give up the rivalry and the conflict.

To my surprise—because I am not unconditionally optimistic about the human species—André and Sylvie saved their family and their marriage.

The Man Whose Brother Had It All

Ahmed, a man of about fifty, came to see me after getting out of a rehab program where he had been placed after attempting suicide by overdosing on sleeping pills.

He comes in accompanied by his wife but wants her to stay in the waiting room and enters my office alone. He looks around, sits down across from me, and says, "I hope you're not recording this conversation and that what we say to each other is strictly confidential."

I reassure him on this count and ask him to explain why he wanted to die.

"I am dying of shame, but I don't think I wanted to die. I truly wanted to sleep and to forget. I adored my father. He was a remarkable man who built an industrial empire in our country and amassed an enormous fortune. Alas, he died twenty years ago. Upon his death, I was very close to my older brother, Omar. We had the same mother and father. I thought that we were going to run the family business together, but over the years, Omar took over everything. He falsified my father's papers and will, and little by little I became his employee. I am the executive director, but he is the president, and he determines my salary. He excluded me from the board of directors and became more and more arrogant, and it's been four years now that we no longer speak or see each other and correspond only by email. My father sent me to France for my studies and then to the U.S., where I got an engineering degree and an MBA. I wanted to do the same thing for my eldest son, who, after high school, asked to be sent to the U.S. for college. I thought that it went without saying and I promised my son who enrolled in a university. And one day my son comes to me and, furious, explains that his American university told him that because he hadn't paid, he was no longer enrolled. I had asked my brother to make the deposit, and I thought that this was completely natural. Well, my brother hadn't done it. Dismayed by what my son told me, I went back to Omar, who gave me to understand that he would not pay for my son's studies and that his decision was final. I was totally humiliated by Omar and I was ashamed and didn't want to tell my son the truth, that is to say my own powerlessness and his uncle's supremacy. It was then that I started not sleeping, not eating, and raging all day long. And finally, I couldn't take it any longer, I wanted to sleep at any cost, and I swallowed all the sleeping pills I could find."

I listen to Ahmed for another hour and I question him. It is clear that his third brain has a precise and designated rival, his brother Omar. His second brain, confronted with the rival's triumph, is producing resentment, shame, and the loss of self-esteem. What kind of mood is it producing? I ask for clarification: "When you look into the bottom of your heart, do you find sadness or anger?"

"Rage and anger."

"And when you examine your brain, your thoughts, are they accelerated or slowed down, do they follow one after the other at high speed and change often, or are they stuck in one place and unfolding slowly?"

"They are accelerated, they come one after the next, they change, they turn in circles. It was also to stop the circus going on in my head that I wanted to sleep."

It is clear that the emotional brain has reacted with an expansive mood of the manic type. This hyperexcitation is sterile and useless, for it bumps up against cultural, social, familial, and religious conventions, in a word against the reality of its impotence. The obvious diagnosis is thwarted mania. Distinguishing this from depression is difficult but crucial, because in this case it is imperative to avoid antidepressants, which would only add fuel to the fire.

And the first brain? It is tempted by paranoia. Ahmed interprets his failure and his brother's success as proof of a conspiracy, given that Omar, according to Ahmed, had certainly falsified papers and bought off notaries, clerks, and judges, but also ministers and politicians.

While listening to him explain the ins and outs of the conspiracy, I am tempted to believe him, because, after all, the result is there: the brother has it all, and he has nothing, except for what his brother is willing to give him. But hasn't the first brain provided him with this coif, this hat that crowns his insufficiencies and perhaps his bad decisions and justifies them?

Concerning the onset and development of paranoia, two sentences from Lacan come back to me. He qualifies the "elementary phenomena" of psychosis as "momentary disorders of perception that are qualified as interpretative because of their obvious analogy with normal interpretation."[1] The other phrase that Lacan offers is in my opinion such a confirmation of my analysis of the three brains and of the primordial and decisive role of interdividual rivalry in psychosis that one might think we were in cahoots, whereas, I repeat, I am not Lacanian: "We have acknowledged as explanatory of the facts of psychosis the dynamic notion of *social tensions*, whose state of equilibrium or rupture normally defines personality in the individual."[2]

I leave it to the reader to meditate on these words and I continue with my interview.

"Have you become emotive and afraid, are there physical phenomena that are bothering you, apart from the insomnia you mentioned?"

"Yes, I have become emotive. I weep easily, especially when I'm talking about all this with someone understanding."

He interrupts himself to shed a few tears and to blow his nose.

"I also experience chills, hot flashes, heart palpitations, and pins and needles in my hands and feet. When I am experiencing strong emotions, there is a lump in my throat that prevents me from speaking."

The second brain is thus tempted to play its own version of the rivalry experienced by the mimetic brain: the somatized emotions are minor forms of hysterical neurosis, of conversion reactions, which as we have seen is the neurotic way of expressing rivalry.

From a nosological point of view, this patient thus has three pathologies: a form of paranoiac psychosis (first brain); a somatic expression of the hysterical type with a conversion reaction (second brain); and a mood disorder of the manic type (second brain). To all of this is now added an addiction to benzodiazepines and to sleeping pills because of a tenacious insomnia, which is characteristic of the manic state. All of these pathologies express and translate the extreme adversarial tension of the interdividual rapport with his brother, whose unfavorable evolution (from Ahmed's point of view) is clothed by the second brain in feelings, emotions, and conversions and hyperthymia, and coiffed by the first brain in interpretations and rationalizations.

I explain to Ahmed that given the number of targets that the various necessary medications are going to aim at and the complexity of the psychotherapy that has to be set up, it would be beneficial for him to spend some time in a clinic. He finds a means of escape and his first brain provides a new avoidance strategy.

"I have decided to go on a pilgrimage to Mecca and to ask Allah for help in my misfortune. If he doesn't bring me a cure, I will come back to see you on my return."

He never came back.

A Dangerous Liaison

Albert, a spry septuagenarian, enters my office and announces that he wants to see me on the insistent advice of his ex-girlfriend, but that he doesn't see what I can do for him! I invite him to sit down and, after telling him that I'm not certain I can help him either, I ask him a "naive" introductory question: "Your girlfriend, you say?"

"Yes, she was my girlfriend for six years, but I left her a year ago. She has to start her life. She's thirty-seven, and she should get married and have kids."

"That's why you left her? For altruistic reasons?"

"You have a good sense of humor. The answer is yes and no. In fact, I left her because I fell in love with one of my secretaries, a woman of twenty-six who is younger and more attractive."

"Bravo! But I don't see the problem."

"The problem is that I'm going crazy. This young woman, Élodie, initially rebuffed me, but nevertheless accepted at first dinners and lunches, and then little by little some presents: clothes, handbags, jewelry. After three months I was madly in love. I thought she was beautiful, elegant, refined, intelligent, and very witty. She was so nice with me that I ended up persuading myself that she loved me. One day, I offered to take her on a weekend trip to Venice and, to my amazement, she agreed. The weekend was dazzling. We made love and I have never been so happy. After that first weekend, we went on many trips, getaways to the most romantic places. And at the same time she was working for me and we understood each other. More and more in love, I gave her a car, and then an apartment where I came to see her every evening. However, little by little, I saw less of her during the day. In the evening, I found her to be as amorous, tender, and adorable as ever. During the day she kept inventing new excuses for longer and longer absences. So I decided to test her: I told Élodie that I had an urgent business trip and I went away for three days. I hired a private detective to stake out her apartment, and he informed me that she was coming home with a handsome young man who never left until just before sunrise. Back in Paris, I reproached her vehemently and revealed to her that I knew that she was receiving a young man in my absence. She denied it at first, pretending that it was a friend she had helped out by letting him sleep on the couch. I told her that I was suffering, that I loved her, that I was jealous, and that I didn't believe her. Then she exploded and changed before my eyes: 'Have you looked at yourself? I've given you my youth. I am free and nobody will deprive me of my freedom. I don't owe anything to anyone.' And she added a sentence that was the coup de grâce and that brought me to you, because ever since I can't sleep, I have nightmares when I do drift off from time to time, and I am completely overwhelmed: 'What are you complaining about? I am faithful to you since I see you every day and I love you. But I will never give up my freedom, and you should never expect exclusivity from me!' I concluded from this that she considered herself free to have one or several lovers so long as she reserved her nights for me. She confirmed this. I was stunned and now I am lost. I still love her and I don't know what to do anymore. I don't believe her anymore, I don't believe anything. I am anxious, sad, life no longer has any meaning, and I've started to drink too much into the bargain. What do you think about all of this?"

I understand that Élodie has plunged Albert's third brain into the most utter confusion: he knows that he has rivals, but he doesn't know them, and meanwhile Élodie is trying to persuade him that he doesn't have any rivals since she is faithful to him! The first brain reveals itself to be incapable of explaining the situation. No suspicion can be formed because everything is so clear and Élodie acknowledges the facts. The first brain has no excuse to provide; it is confused and disoriented, and the only thing it has managed to do is to pull on the emergency cord and insist that Albert get help. As for the second brain, it is devoured by jealousy, by anxiety, by contradictory emotions, and it is secreting a labile mood: now depressive and practically suicidal, and now euphoric and optimistic at the slightest cuddle from Élodie.

"I am lost," says Albert. "I am confused. I no longer recognize myself. This has never happened to me before. It's as if the world has turned upside down and no longer has any meaning. It's so strange."

Amorous passion brought to a point of incandescence by rivalry—especially when there is no one single rival but many potential rivals—ends up presenting clinically in the form of an acute psychosis with a loss of reference points, a loss of meaning, and a feeling of strangeness. It is therefore urgent to restructure the rivalry and to stabilize the position of the third brain. To achieve this, one must address the third brain and mobilize it, while calming the anxiety and disorderly emotions of the second by means of some light tranquilizers. As with all acute psychoses, one has to take immediate action. Among young people (both men and women), this psychosis can become chronic, with the third brain seeing the beloved as impossible to attain, that is to say as an obstacle. Among older people, like Albert, the risk is a collapse of the second brain with severe depression and a risk of suicide.

To clarify and fix his rivalry, I speak to Albert with a certain solemnity.

"You are in danger. And it's a mortal danger. If you don't react, Élodie will end up destroying you. You are a powerful CEO and a successful businessman. You must act remorselessly, as you often have, I am sure, in your business deals. The decision is yours, but my duty is to warn you and to advise you to break off all contact with Élodie, whom you must see clearly as an enemy capable of massacring you."

Albert sits up in his chair, looks at me for a long time, and nods his head slowly. Then he says to me: "Your advice is clear. But I am going to think about it. I am going to suffer if I make this choice. Will you be able to help me?"

"I will help you if you make the decision and you cut all ties with Élodie. You are going to suffer, but I will be there and I will support you. The medications can also help soothe and relax you and enable you to sleep."

The clear identification of the rival helped Albert's mimetic brain and charged his rational brain with taking the necessary measures. The treatment helped his emotional brain to bear up, and I must say that I admired his courage and determination.

A Salutary "Betrayal"

Finally, to illustrate the fundamental role of identification—a therapeutic process whose exacerbation can lead to a veritable passion for the model—I am going to tell you the story of Suzanne.

She was sixty-three when she came for a consultation, and she told me that, forty years earlier, she was working in a company in the same office as Sabine. They became friends, and then each became the other's best friend. Sabine met a young German, Wolfgang, who was in training at the company, and married him. She went to live with him in Munich and they had three boys. Suzanne continued to work in Paris but went regularly to Munich to see Sabine; and Sabine also came to see her in Paris and they often went on vacation together. Sabine's children grew up and grew attached to their mother's best friend over the course of many visits. Suzanne didn't get married and had many liaisons. For some years, however, she had had a steady boyfriend, who was married, and who for that reason didn't hinder her precious freedom!

Suzanne was depressed when she came to see me, because three years earlier Sabine had succumbed to cancer. Suzanne had been at her side until the very end, and then, gradually, refused to accept her death. This refusal was expressed through her identification with Sabine. She went as often as possible to Munich laden with gifts for Sabine's sons, who were now adults, and during her stays she took charge of the errands, the cooking, the dishes, and so on. Quite naturally, she had become attached to Wolfgang and had begun to spend vacations with him and the children. To complete her identification with Sabine, she was now entertaining the idea of marrying Wolfgang. And she wanted to have my opinion!

Instead of giving it to her, which was not my place, I asked her what the status of her relationship with her boyfriend was. She told me that he was of Russian extraction, and that he had a name that was impossible to pronounce, so she called him Zag. He was, according to her, as easygoing and light as Wolfgang was heavy and difficult. She confessed to me that she never spoke about Zag to Wolfgang, but that Zag knew of Wolfgang's existence because Sabine and Suzanne had been speaking about him in Zag's presence for years. I made no comment, but I understood that Suzanne wanted to forget Zag and to attach herself to Wolfgang for the sole purpose of prolonging and resuscitating—by means of this identification—the life of Sabine. But she kept running up against Wolfgang's difficult nature and the fact that she was not in love with him.

Suzanne came back to see me a few months later, looking radiant. Noting my surprise, she explained to me that Zag, to whom she had recounted her stays in Munich and her difficulty in dealing with Wolfgang, advised her to speak to his sons so that all of them could advise Wolfgang to consult with a psychiatrist, as his bad mood, in Zag's view, was probably due to his inability to get over Sabine's death.

Suzanne took advantage of her next stay in Munich to speak with one of the three sons, the one she was closest to. She told him that the idea of consulting a psychiatrist had come from Zag, but that he absolutely mustn't say anything to his father and should instead present the idea as his own. That very evening, at dinner, the young man, Otto, broached the subject, declaring right from the start to his father that Suzanne thought he should consult a psychiatrist and that the idea was Zag's! Shocked and astonished by Otto's "betrayal," Suzanne remained silent for a moment. Then she reacted and recounted in detail the discussions she had had with her boyfriend on this subject, and his opinion, which she found worthwhile, insisting that Wolfgang should see a psychiatrist. Wolfgang shut down, obviously vexed, and said nothing in response.

In fact, young Otto had exploded Suzanne's total identification with Sabine. With his attitude, he had clearly signaled to Suzanne that she wasn't his mother, would never be his mother, and that he was entirely on his father's side. This brutal uncoupling of an overly clingy interdividual rapport, which was becoming pathological, enabled Suzanne to get back in touch with reality.

That evening, Suzanne had suddenly experienced an immense feeling of well-being. She said that she felt liberated, happy to have spoken of Zag, to have told Wolfgang what she thought of his bad moods. In fact, she had become aware of the role she was trying to take on—that of the dead and gone Sabine. And it was Otto who had triggered this sudden realization by speaking to his father about Zag, and failing to respect the promise he had made to Suzanne. By doing that, Otto refused to make room for Suzanne in the family, sending her back to her life with Zag and dismissing her from her host family in Munich. Shocked at first, Suzanne then realized that the obligation she had imposed on herself was too weighty, and that day she got over Sabine's death. She was "healed."

Conclusion

When I was a child, sitting at the back of the classroom, I looked at the blackboard and I couldn't make out anything. I wondered why the others saw what the teacher was writing and I didn't. My parents ended up realizing that I had a vision problem; they bought me glasses, and I could finally see as well as my classmates. I had a similar experience when I began to study psychology and psychiatry. I was given psychoanalytic glasses, but I didn't see anything much better. I was given cognitivist glasses, and they didn't help me any more than the psychoanalytic ones. That is why, when René Girard gave me mimetic glasses, I was relieved to be able at last to see symptoms and syndromes clearly, and to be able to discern the mimetic patterns in human behavior. This led me to adopt a very modest approach to psychotherapy, since I knew that the mechanisms that I have described throughout this book are implacable and that the only way to free oneself from them is to face up to them and to recognize them in all humility.

Just as God created man in his image, animating the clay with otherness, so do human beings engender one another mutually, not only on the genetic level but also on the psychological level, the self being filled with and saturated by otherness throughout its history and constituted as a patchwork of all integrated others. The stubborn denial of this reality corresponds to original sin which is expressed at N in a neurotic fashion and at N' in a psychotic fashion. Psychotherapy must take this reality into account and integrate the very essence of this rebellion against the real that is commonly called "mental illness." The practice of initiation must bring about the gradual metamorphosis that leads to wisdom through the successive and never entirely completed recognition of the otherness of which we are made up.

The essence of the human condition is the "management" of this unbearable otherness, with which we are saturated and by which we are encumbered. Here, too, Lacan came very close to this reality when he cites Paul Valéry: "There is in each of us . . . contradictory but indivisible energies. One is the eternal movement of a big *positive electron* that repeats a profound, monotonous phrase: *There is only me. There is only me, me, me.* . . . As for the little radically *negative* electron, it cries at an ultrahigh pitch: . . . *Yes, but there is someone . . . yes, but there is someone . . . someone, someone, someone.* And someone else!"[1]

Human beings look at the world in a subjective way that is shaped by mimetic desire, which is an illusion. Desire modifies the very aspect of objects and in chapter 3 of Genesis, we see Eve's perception of the tree change as she is gradually penetrated by the mimetic desire represented by the serpent. In *Things Hidden since the Foundation of the World*, Girard and I even considered the totally illusory emergence of an object of desire that we called the "hallucinatory psychosis of desire." This hallucinatory psychosis can indeed literally make you see objects that don't exist, or, much more often, make you see objects in a distorted light, make you think the moon is made of green cheese, or, like Don Quixote, that an old inn in front of which two prostitutes are chatting is in fact a chateau graced by two noble ladies exchanging immortal remarks. Seeing reality as it really is thus requires an effort, but it's a necessity if we want to solve problems and leave illusion behind.

As I have often emphasized, I think that the opposite of madness is not mental health. The opposite of madness is wisdom. And wisdom is the long transformative—that is to say initiatory—process by which each of us can gradually recognize the mimetic mechanisms of which one is the plaything, overcome the mimetic rivalries of which one is the prisoner, and avoid even the most scandalizing and staggering mimetic obstacles, so as to move toward a situation of calm, harmony, and peace inside oneself and between oneself and others.

I think that attaining wisdom, the type of wisdom that I am talking about, is the objective of Socrates, Buddha, Christ, Krishnamurti, the Dalai Lama, and all the great sages. What I am seeking to contribute is a scientific framework that makes it possible to orient oneself along this initiatory path and that offers the therapist constant feedback about what he is doing, and enables him or her to follow along with the gradual harmonization of the three brain functions all the way until—if possible—their final harmony.

Certain techniques or teachings call especially upon the first brain, like those of Socrates or Krishnamurti. Others are based primarily on a mastery or pacification of the second brain, like those of Christ ("Love one another"), Milton Erickson, J. H. Schulz (founder of "autogenic training," a relaxation technique), Gurdjieff, and Buddha. I have the feeling that awareness of mimetic mechanisms and of the primordial and automatic action of the mimetic brain as well as its workings is cruelly lacking in the methodology that has to be put in place in order to arrive at wisdom. For the great sages, whose teaching bears essentially on the first or second brain, consider (without saying so) that the third, which they do not mention, could be influenced by the other two sufficiently to attain wisdom.

It is clear that few people have been able to take full advantage of the teachings of these great sages, and I think that their number can be significantly increased thanks to scientific awareness of the existence and workings of the third brain as well as of its dialectical and constant interaction with the other two. Simply stating the goal to be achieved and the innumerable failures of all the teachings offered by the great sages shows us how difficult the undertaking is. This echoes Christ's words: "Many are called but few are chosen." This also echoes the fact that the dead ends each of us can wander into are innumerable, while there is but one path that leads to wisdom. It is thus with the greatest modesty—the greatest patience, the greatest benevolence, the greatest indulgence—that the therapist, now conscious of the global reality of the psychic apparatus, must go in search of each patient on the dead-end path where he or she has lost the way, to attempt to guide the patient by means of every initiatory process at his disposal, addressing now the first, now the second, and now the third brain according to which target seems best to him.

Psychoactive drugs can also be of help. Neuroleptics act on the hallucinations and delusions of the first brain as well as on the aggression and violence of the second. Tranquilizers act on the second brain's anxiety and emotivity. Antidepressants act essentially on the second brain with the hypothalamic-pituitary-adrenal axis as their primary target; the same axis is the target of normothymics like lithium tasked with stabilizing mood and preventing manic or depressive bipolar swerves. It is clear that medications capable of acting on the third brain and of influencing mirror neurons to promote empathy and a relationship to the model as such have not yet been created, even if oxytocin seems to have an effect along these lines. But pharmacological research is only in its infancy in this domain.

As we saw with respect to the constitution of the self, forgetting is healthy, but wisdom lies in recognition. If the subject is in a state of forgetting, he or she is in a state of normalcy. But those who are incapable of staying in a normal state are obliged to experience the ordeal of rivalry. For them, recognition is the supreme goal of all psychology and of all initiatory practices.

In our day practicing psychiatry is going to become more and more difficult for young psychiatrists, first of all because semiology and nosology vary so much from school to school. The DSM-IV or DSM-V and the CIM-10 may appear overly simplified or even inhuman to them. These classifications, which are "statistical" by their own admission, are insufficient. Moreover, psychoanalysis may not have an answer for all their questions and is less and less likely to resonate with their patients. Cognitivism, while very useful for addressing certain symptoms, namely those related to phobias and obsessions, may appear inefficient for dealing with other pathologies. Finally, the use of medication is becoming more and more delicate and, I would add, dangerous—not for the patients but for the doctors—for the proliferation of rules and recommendations, and the attention being given to side effects, is making patients more suspicious, vindictive, and litigious than they are eager for healing. After the euphoria that accompanied the discovery of various psychoactive drugs in the 1960s, 1970s, and 1980s, we have reached the point where people mistrust and disapprove of medications. Today patients and the public in general are more afraid of the treatment than they are of the illness itself, and with this fear of treatments an old idea is making its way insidiously back into our culture: the denial of mental illness.

Young psychiatrists are thus being confronted with new and very serious difficulties that must be addressed first of all by an increase in their familiarity with human realities, a deepening of not only their biological and

biochemical knowledge but also and above all their philosophical, psychological, anthropological, and even ethnological knowledge, given the variety
of patients they will have to treat and the cultural differences among them.
They will also have to work on themselves to conquer their own passions,
that is to say the mimetic mechanisms of which they are the playthings. They
will only be able to help their patients by being slightly ahead of them on the
path to wisdom and renunciation of rivalry—with the other being taken as a
model and as little as possible as a rival or an obstacle. The wise way of desiring leads to a preference for what one has rather than for what one doesn't
have, and in all circumstances to renouncing comparisons between what one
has and what one's neighbor has.

Thus, psychotherapy will gradually become mimetic not only in its
technique of diagnosing the processes at work in the patient, but also to
the extent that the psychotherapist's wisdom will serve as a model for the
patient. The psychotherapeutic mimetician will then be able to establish an
interdividual rapport with the patient that is as free as possible from rivalry
and from the *skandalon*,[2] a rapport that will inspire the patient in his relationships with others.

◆   ◆   ◆

All of this has been said already by Voltaire in a humorous way, and in *Candide*
we find many of the elements addressed in this book. The great "novelistic"
writers are better psychologists than are the officially recognized psychologists themselves. That is why in this book I have given as much importance to
the texts we have inherited from great authors as to the clinical cases I have
observed in my practice.

At the end of *Candide*, Voltaire highlights the mimetic mechanism on
two occasions. First by showing the effect of taboo on the reinforcement of
desire: "At the bottom of his heart, Candide had no wish to marry Cunégonde, but the baron's insolence made him determined to go forward with
the marriage."[3] This echoes the reply of Lacan's patient "Aimée" to her family's objections to her marriage: "If I don't take him," she says of her fiancé,
"someone else will."[4]

Then Candide encounters an elderly Turkish gentleman who declares
that he pays no attention to public affairs and lives by selling the fruit he
grows in his garden. "'You must have a vast and magnificent estate,' Candide

said to the Turk. 'I only have about twenty-five acres,' the Turk replied. 'I cultivate it with my children. Work keeps three evils at bay: boredom, vice, and want.'"[5] Candide goes back to his farm, and there he "reflects profoundly" on what the Turk said about work and its positive effects. He finally tells his companions, Pangloss and Martin: "That fine old man seems to have secured himself a better fate than that of the six kings with whom we had the honor to dine."[6]

From a mimetic point of view, Rousseau, as we saw earlier, senses the danger of mimetic rivalry and makes the object responsible for it: in his view it's private property, that is to say the object that is appropriated for oneself, that brings about rivalry and thus social disorder and unhappiness. What he doesn't see is that it is not private property that must be abolished but rivalry. It's not property that creates rivalry but the comparison between that property and what one has oneself, which is smaller or perhaps nonexistent.

Voltaire, on the other hand, understands two things: first, to cultivate one's garden is to refrain from comparing it to other gardens, which may be bigger or smaller; second, working in one's garden is the remedy for sadness, envy, jealousy, and as it happens for basic needs, too. The work that Voltaire speaks about is the same work that I spoke about in my book *Psychopolitics*. The Turk who cultivates his garden doesn't have a "job" and doesn't keep track of his hours. His work in the garden is one and the same as his life. It would be ridiculous for him to think that leaving half his garden unplanted after working thirty-five hours was a guaranteed social right.[7] The day that human beings think of work—not just work on external objects but also work on themselves—as being the very essence of their lives, they will no longer keep track of the hours they devote to this task, for the simple reason that the hours not devoted to it will in some sense be hours of nonlife.

In reality, working on oneself means working on the interdividual rapport, that is to say on one's relationships with others. Daniel Goleman's book *Social Intelligence* resembles this one in many respects, but without having at its disposal the mimetic eyeglasses that have been so precious to me.

In daily clinical experience, one sees retired people who, after having declared for years their eagerness to retire in order to have time to go to the movies, the theater, fishing, and so forth, are depressed because now that they have time, they no longer have any motivation or desire to do all of those amazing things they promised themselves they would do. So they do nothing, and the cases of Alzheimer's proliferate: after doing nothing, one *is* nothing.

Cultivating one's garden, investing one's energy and desire in that undertaking, protects the Turk from two misfortunes expressed earlier by Martin: the "convulsions of apprehension" and the "lethargy of boredom."[8] Work is a way of concentrating on *what one has* instead of concentrating on *what one doesn't have*. It is the remedy to the discontents of humanity and to the traps of mimetic rivalry.

When God put man in the Garden of Eden, he didn't ask him to work the land. He asked of him only the spiritual work that consists in living in the light of God, according to God's will, that light and that will being symbolized by the observance of a simple rule: don't eat from the forbidden tree. But mimetic desire—represented allegorically by the serpent—ends up being stronger, and it thrust man into the world. And there, he is obliged to work the land to make it produce the fruits and vegetables that grew on their own in the light of God in the garden of Eden. Once again, it is mimetic desire that has plunged us into all sorts of problems, pain, and misfortune. And it is only by working on oneself to undermine the mimetic mechanisms that make us their puppets that we can survive the human condition characterized by Martin, described as catastrophic by Rousseau, and to which Voltaire's Turk brings a remedy: just worry about your own garden. Don't worry about what's happening in Constantinople, and make your work your life, and your life your work.

The image that Voltaire gives at the end of the book is that of a smoothly functioning society, after its members have experienced all the setbacks that mimetic rivalries create for human beings: after sampling one by one all the chatoyant illusions of mimetic desire, after going through all the snubs and disillusionments of rivalry, one finally decides to cultivate one's garden, which is a modest objective, it is true, but one that offers a happiness that mimetic and antagonistic undertakings proved incapable of providing.

Let us conclude, then, that the lion's share of the work must not be done on the object but on oneself. And work on oneself, like all work, begins with waking up. One cannot work on anything at all when one is sleeping. Waking up is thus the first condition of work properly understood, of work on oneself, and thus the first step on the path to happiness.

◆ ◆ ◆

At certain moments, we recognize that we have taken someone for a model and that, having taken him as such, we have envied him and that he has become our rival. This recognition is therapeutic. We find such a moment of recognition in Lionel Duroy's novel *Le chagrin*: "Gradually, over the years, I became closer to Frédéric. I was going to write that Frédéric had become my model, but no, as far back as I can remember, Frédéric has *always* been my model. . . . I would have been willing to jump out a window without a parachute to earn his esteem. . . . As far back as I can remember, my older brother is a lord whose heart I must conquer to attain pleasure, happiness, life itself. . . . I envy his cholera, which made him a slender, fragile child whom our mother was constantly looking at with an anxious gaze."[9] Later on in the novel, Duroy expresses the temptation to rivalry, which is quickly repressed, and the return to the other as model: "Despite a certain complicity between us, Frédéric has in my eyes a great capacity for doing evil. And yet he remains my model, the prince whose esteem and friendship I continue to seek out."[10]

The older brother is his model and inspires him. The narrator wants to become a writer "because my older brother told me one day that he wanted to be a writer."[11] Later, the narrator has just quarreled with his girlfriend Agnès and he suddenly realizes that he is repeating and rehearsing the behavior of his father (Toto) and mother: "I am relieved, we are reconciled. I think to myself that from now on I will be wary of my anger. I am not Toto and Agnès is not my mother, she doesn't want to humiliate or destroy me. . . . I want . . . us to remember this for the rest of our lives, to recall that on that day we left our childhood behind and invented a way of loving one another that was different from that of our parents."[12]

This kind of episode is what happens all the time over the course of a psychotherapy and is called a *prise de conscience*, or "sudden realization." It is very often the abrupt recognition of someone else's behavior (parents, brothers, sisters, teachers) in our own. The reaction may be a feeling of affection or one of rejection, as in the case I just cited, in which the model becomes not a rival but an antimodel: in becoming aware of the model's influence (of his suggestion), one rejects that influence, refusing to imitate, and does the reverse to avoid all resemblance with the despised or detested behavior.

◆ ◆ ◆

Having arrived at this stage in our reflections, we must recognize that up until now, to understand and communicate with each other, we—you who are reading the book and I who have written it—have used only our first brain, our cognitive brain. Perhaps some readers have nevertheless felt themselves to be empathically in tune with the ideas expressed herein. Perhaps others have had negative reactions. These reactions of the third brain triggered positive feelings in the first group's second brain and hostile ones in the second group's.

If all of this is without much importance here, in the therapeutic context it is highly important. We have seen that to address the first brain, we have at our disposal a huge variety of psychotherapies, notably those qualified as "cognitive." Psychoanalysis also addresses the first brain by trying to bring "repressed" memories to the surface using techniques of free association and dream interpretation. Some techniques, like Janov's "primal scream therapy," mobilize essentially the second brain.

On the other hand, to speak to the third brain, the only tools we have at our disposal are ourselves. The discovery of mirror neurons has taught us the decisive importance of reciprocity, such that smiles, amiability, and politeness generally bring about a mirrored attitude that makes social life possible if not agreeable. Laughter is a particular and very interesting form of interdividual rapport. It can be cynical, ironic, mocking, and aggressive, and entail the humiliation and hatred of the one who is its target. But shared laughter and laughing fits strengthen the bonds of friendship. And I have noticed that if during a psychotherapy I manage to get the patient laughing along with me, it always means real progress.

And yet the only form of psychotherapy that is solely and directly addressed to the third brain is classical hypnosis, from which numerous forms of psychotherapy have been derived: Schultz's autogenic training, directed daydreaming, EMDR (eye movement desensitization reprocessing), Ericksonian hypnosis, and so forth.

The hypnotic rapport is a special interdividual rapport in which the imitation and suggestion vectors—which in normal relations have a cinematic quality and are constantly moving back and forth—have been fixed in place: in the suggestion position from the hypnotist to the patient and in the imitation position from the patient to the hypnotist. Stabilizing these two vectors brings about the substitution of the patient's desire d and its replacement and guidance by the hypnotist's desire D. Thus does d disappear, and with it the self dissolves and vanishes. This is the period of lethargy. The dissolution of self s can be compared to a "ritual agony." It is initiatory in that it reveals the genesis of the self elicited by desire and its disappearance when the desire that produces it is erased and is replaced by another desire. The vanishing of self s, the "habitual self," is equivalent to a "ritual death."

In the remaining stages of the trance, desire D, the hypnotist's desire, produces a new self, s', in the patient. And this self "wakes up"—this is the somnambulistic state. The newly formed self, s', engendered by desire D, has a new consciousness, memory, and "personality." Newly formed, self s' can remember self s, the habitual self, because it is posterior to that habitual self in physical time. After the trance, self s, when it "comes back to itself" and wakes up, will obviously be unable to remember self s', whose existence comes after it and is thus in the future with respect to it. At the most, it would be able to "foresee it," which is impossible.

What is remarkable is that in the course of this process, the process of the hypnotic trance, self s lives through a properly initiatory experience of ritual agony, death, and resurrection. At nodal points N and N' it experiences the lived reality of physical time, the priority of desire D over the desire d that it constitutes, that is to say that it experiences in real time the creation of the self by desire, the birth of a new self-of-desire. This experience is an initiatory one that is in my opinion extremely valuable for the third brain, because it reveals the workings of the third brain, the genesis of its "own" desire and that of its "own" self. The hypnotic experience could thus be used as an initiation into the truth and reality of the interdividual rapport and, by this token, it could repair at N and N' the damage caused by taking the model as a rival or an obstacle, since throughout this initiatory experience the model remains such.

Have we rediscovered the meaning of psychotherapy? Yes and no. Pierre Janet invites us to be wary of "somnambulistic passion," the patient's attachment to his or her hypnotist: he can't let go and wishes to live in a sort of permanent state of hypnosis, tormented by what Janet calls the "need for direction." This "need for direction" represents the patient's dependency on the therapist, which is the opposite of the goal that all forms of psychiatry and psychotherapy are aiming at: the patient's independence and conquest of freedom.

When psychotherapy goes in the direction of making the patient dependent on the therapist, it follows the dangerous path of least resistance. And it bears within itself in germ all autocratic and theocratic abuses. Let us pay heed to Stefan Zweig: "Out of lassitude when faced with the frightening number of problems, the complexity, and the difficulty of life, the vast majority of human beings aspire to a mechanization of the world, a definitive order ... that would enable them to avoid the hard work of thinking. It is this messianic aspiration ... that constitutes the veritable unrest preparing the way for all social and religious prophets."[13] Prophets who are transformed into theocratic, political, or military dictators, like some hypnotists or therapists, seduced by their patients' "need for direction," are transformed into gurus and slide down a slippery slope that leads to abuses of power. On a social or political scale, Stefan Zweig calls this kind of psychotherapist a "redeemer": "Millions of individuals are ready, as if by enchantment, to let themselves get taken in ... and the more this redeemer requires of them, the more they are ready to give. That which, only yesterday, was their supreme happiness, freedom, they give up out of love for him, to let themselves be guided easily along."[14]

Thus a single mechanism is at work causing the third brain to take the model as an absolute model, a unique model, or as Freud would say, an "ego ideal," both in the patient who has become dependent on his guru and enslaved by his somnambulistic passion, and in the masses, the thousands or millions of brains of a people who have become dependent on their political, military, or religious leader. The fact that the two mechanisms are identical was pointed out by Freud when he wrote that "the hypnotic rapport is 'a crowd with only two members." And in both cases, what leads to the abandonment of freedom and to dependency is the "need for direction," that is to say the path of least resistance for mimetic desire, which delegates the choice not of a single object but of *all its objects* to a single, absolute model.

This means that, as in all initiatory processes, hypnosis—if one wishes to use it as an initiation into psychological reality, as a revelation of the mimetic essence of the interdividual rapport, as a lived experience of the model that remains such—requires long preparation. The sort of hypnosis you find performed in carnivals and music halls or on television is without object. As for medical hypnosis, it is not initiatory because it is not preceded by a long preparation, nor is it followed by a revelatory commentary on what has been experienced on the metapsychological level.

If it is used as an initiatory process, hypnosis must be the outcome of a psychotherapy that gradually builds up to that lived experience. In the course of this psychotherapy, which in reality is an evolution along the road to wisdom, a fundamental stage in the experience is "letting go." Many great sages have emphasized the importance of meditation as a means of concentrating on the present. "The Power of Now"[15] is of great importance. Concentrating on the present moment is a temporal letting go that consists in abandoning the past with its cortege of remorse and regrets, forgetting the future with its cortege of expectations, hopes, and anxieties, and living intensely in the only time that we always let slip through our fingers, and to which we do not pay attention: the present time, the here and now, where I am living, breathing, looking, contemplating, where I am!

Then the hypnotic experience that enables us to experience letting go in the present, abandoning ourselves to the model who dissolves the self and resurrects it in another form, can be fully profitable and initiatory, and thus therapeutic in the noblest sense of the term: an initiation into wisdom, a clear vision of reality, a privileged moment where illusions vanish.

◆   ◆   ◆

Thus, throughout our lives, initiatory experiences organized by culture, as in the case of adorcism, or organized by interaction with a teacher, would enable us to exorcise that fear of death which is the great problem that obsesses human beings, whether they realize it or not. The idea that one died yesterday in order to live today and that today will die to make way for tomorrow would make it possible for human beings to be reassured and to grow, the fundamental lesson of initiatory rites being that death brings about a different but more advanced and better state of being, that the caterpillar dies so that it can be transformed into a butterfly, and that it is not the adolescent who dies but adolescence.

The problem of psychiatry is that it is dealing with patients who, to be sure, are afraid of death just like everyone, but who are also afraid of life. For the active denial of the otherness of desire at N and the claim to the precedence of desire over the other's desire at N' constitute an active refusal to recognize reality, that is to say life. This misrecognition, whether it takes a neurotic or psychotic form, is a process that consists in going back in time and refusing, despite all the evidence, the reality of physical time, and thus, in a certain sense, in embarking on an undertaking that is the inverse of the initiatory one. The psychotherapy that could be called "mimetic" or "interdividual" aims to reassure patients and to lead them to commit first to the here and now and then to the decision to move forward. To do this, one must give them confidence in life so that they can be buoyed up and pushed along by the vital current rather than wasting their energy trying to swim against it.

It then becomes abundantly clear that all recognition of mimetic reality and of the otherness of the desire that constitutes us will be a step on the path to healing, a soothing of existential anxiety, and of the fear of life. Obviously, the psychotherapists who want to adopt this approach must lead the way and serve as models and examples by recognizing the otherness of their own desire and the priority of the other's desire over the desire that constitutes the self and brings it into existence. These qualities should enable them to show patients that it is possible to live without being afraid. Freud recommended that psychoanalysts should themselves be analyzed. This is a very wise recommendation because to lead the way one must be more advanced than the people who are following along behind. I think that the mimetic psychotherapist must have gone through enough initiatory experiences and have evolved sufficiently in recognizing the otherness that constitutes him or her to be as far along as possible on the path of wisdom, before guiding the patient along it in turn.

This book's ambition is to make a scientific—as opposed to interpretative or subjective—psychological and psychiatric contribution to the initiatory process so as to make this undertaking more accessible to psychotherapists, first of all, and then to patients.

Notes

Preface

1. I would like to draw the reader's attention to the fact that in French, "conscience" and "consciousness," which are distinct notions in English, are denoted by the same word: *la conscience*. This has led to endless debates among philosophers and psychologists in France.

2. Broadly speaking, Freud's first topic distinguishes the conscious mind from the unconscious. The second postulates the existence of three entities: the id, the ego, and the superego.

3. Later translated into English by Eugene Webb under the title *The Puppet of Desire*.

Introduction

1. An abundant popular literature on the subject has appeared over the last decade, among which Christian Keysers's *The Empathic Brain*, Marco Iacoboni's *Mirroring People*, and, more recently, in a skeptical vein, Gregory Hickok's *The Myth of Mirror Neurons*.

2. Vittorio Gallese, "The Shared Manifold Hypothesis," *Journal of Consciousness Studies* 8, nos. 5–7 (2001): 33–50.

3. Gallese, "The Shared Manifold Hypothesis," 46.

4. Gallese, "The Shared Manifold Hypothesis," 47.

5. Stephen A. Mitchell, *Relational Concepts in Psychoanalysis: An Integration* (Cambridge: Harvard University Press, 1988).

6. Mitchell, *Relational Concepts*, 21.

Chapter 1. Contagious Desire

1. René Girard, *Deceit, Desire, and the Novel*, trans. Yvonne Freccero (Baltimore: Johns Hopkins University Press) 1965, 3.

2. Girard, *Deceit, Desire*, 5.

3. Girard, *Deceit, Desire*, 6.

4. Girard, *Deceit, Desire*, 10–11.

5. Girard, *Deceit, Desire*, 12.

6. Girard, *Deceit, Desire*, 17.

7. Girard, *Deceit, Desire*, 23.

8. Girard, *Deceit, Desire*, 33.

9. Girard, *Deceit, Desire*, 42.

10. Marcel Proust, *The Captive*, cited in Girard, *Deceit, Desire*, 48.

11. Girard, *Deceit, Desire*, 53.

12. Girard, *Deceit, Desire*, 53.

13. Girard, *Deceit, Desire*, 53.

14. In this book I use the word "initiation" in its more anthropological meaning—a mechanism by which metamorphosis is accomplished, for instance a rite of passage that transforms the child into an adult. In the Girardian sense of Proustian initiation, the Proustian snob just wants to imitate the model and be integrated into his world, hoping to acquire or share its prestige. This is at once very similar to and quite different from the metamorphosis of the rite of passage, which entails a transformation of the self toward a higher level of being.

15. Girard, *Deceit, Desire*, 83.

16. Girard, *Deceit, Desire*, 85.

17. Girard, *Deceit, Desire*, 87–88.

18. Girard, *Deceit, Desire*, 88–89.

19. Girard, *Deceit, Desire*, 89.

20. Girard, *Deceit, Desire*, 90.

21. Girard, *Deceit, Desire*, 103.

22. Girard, *Deceit, Desire*, 105.

23. Girard, *Deceit, Desire*, 107.

24. Girard, *Deceit, Desire*, 109.

25. Sections of this book (the cases studies pertaining to Mrs. H, Micheline, and Mademoiselle L.) were adapted from my article "Desire Is Mimetic: A Clinical Approach," which was published in *Contagion: Journal of Violence, Mimesis, and Culture* 3, no. 1 (1996): 43–49.

Chapter 2. The Precursors

1. See on this subject Wolfgang Palaver's *René Girard's Mimetic Theory* (East Lansing: Michigan State University Press, 2013), in which Palaver offers examples of precursors from the religious and philosophical tradition, including Sartre, Kierkegaard, Augustine, Spinoza, and Hobbes.

2. *The Collected Works of Spinoza*, vol. 1, trans. Edwin Curley (Princeton: Princeton University Press, 1985).

3. Spinoza, *Collected Works*, 512.

4. Spinoza, *Collected Works*, 512.

5. Spinoza, *Collected Works*, 513.

6. Spinoza, *Collected Works*, 514.

7. Spinoza, *Collected Works*, 521.

8. Spinoza, *Collected Works*, 493.

9. Spinoza, *Collected Works*, 589–590 (my emphasis).

10. Spinoza, *Collected Works*, 513.

11. Spinoza, *Collected Works*, 512.

12. Spinoza, *Collected Works*, 562.

13. Sigmund Freud, *Contributions to the Psychology of Erotic Life*, in Sigmund Freud, *The Psychology of Love*, trans. Shaun Whiteside (New York: Penguin, 2007), 239–278.

14. See Sigmund Freud, *Psychologie de la vie amoureuse* (Paris: Payot & Rivages, 2010), preface by Robert Neuburger, 11–12.

15. Freud, *Contributions*, 242.

16. Freud, *Contributions*, 242.

17. Freud, *Contributions*, 243.

18. Freud, *Contributions*, 243.

19. Freud, *Contributions*, 243.

20. Freud, *Contributions*, 243.

21. Freud, *Contributions*, 244.

22. Freud, *Contributions*, 244.

23. It is clear that in the text that I have just cited, Freud speaks only of the man's desire for the woman and of the mechanisms by which the man chooses his "sexual object." There is no need to specify that one could say the same thing about the woman and the choice of the "sexual object" of her desire.

Chapter 3. Some Contemporaries

1. Andrew N. Meltzoff and M. Keith Moore, "Explaining Facial Imitation: A Theoretical Model," *Early Development and Parenting* 6 (1997): 179–192.

2. Andrew N. Meltzoff and M. Keith Moore, "Imitation of Facial and Manual Gestures by Human Neonates," *Science* 198, no. 4312 (October 7, 1977): 75–78.

3. These remarks were made a meeting of the so-called Garrels Group, an interdisciplinary group brought together to explore the link between René Girard's theory of mimetic desire and violent cultural origins and research in the empirical sciences. The meetings, sponsored by the Templeton Foundation, eventually gave rise to an edited volume: *Mimesis and Science: Empirical Research on Imitation and the Mimetic Theory of Culture and Religion*, ed. Scott Garrels (East Lansing: Michigan State University Press, 2011).

4. Andrew N. Meltzoff and Rechele Brooks, "Eyes Wide Shut: The Importance of Eyes in Infant Gaze-Following and Understanding Other Minds," in *Gaze-Following: Its Development and Significance*, ed. Ross Flom, Kang Lee, and Darwin Muir (Mahwah, NJ: Erlbaum), 217–241.

5. Andrew Meltzoff, "Imitation, Gaze, and Intentions," in Garrels, *Mimesis and Science*, 55–74.

6. Meltzoff, "Imitation, Gaze, and Intentions," 69.

7. This comment is cited by Ben Thomas in his guest blog, "What's So Special about Mirror Neurons?" *Scientific American*, November 6, 2012.

8. In recent years, new means—such as a technique called "continuous theta-burst stimulation," which impairs brain areas to determine which areas perform which functions—have been used to observe the activity of mirror neurons.

9. Giacomo Rizzolatti and Corrado Sinigaglia, *Les neurones miroirs*, trans. Marilène Raiola (Paris: Odile Jacob, 2008), 10.

10. Rizzolatti and Sinigaglia, *Les neurones miroirs*, 11.

11. Vittorio Gallese, "The Two Sides of Mimesis," in Garrels, *Mimesis and Science*, 96.

12. Vittorio Gallese et al., "Action Recognition in the Premotor Cortex," *Brain* 119 (1996): 593–609.

13. Vittorio Gallese, "The Intentional Attunement Hypothesis: The Mirror Neuron System and Its Role in Interpersonal Relations," in *Biomimetic Neural Learning for Intelligent Robots*, ed. Stefan Wermter, Günther Palm, and Mark Elshaw (Berlin: Springer, 2005), 19–30..

14. See Patrice van Eersel, *Votre cerveau n'a pas fini de vous étonner* (Paris: Albin Michel, 2012).

15. Christian Keysers, *The Empathic Brain: How the Discovery of Mirror Neurons Changes Our Understanding of Human Nature* (n.p.: Social Brain Press, 2011).

16. Marco Iacoboni, *Mirroring People: The Science of Empathy and How We Connect with Others* (New York: Farrar, Straus and Giroux, 2008), 106–109.

17. Vittorio Gallese, "Embodied Simulation: From Neurons to Phenomenal Experience," *Phenomenology and the Cognitive Sciences* 4 (2005): 23–48.

18. See Aristotle, *Poetics*, IV, 2.

Chapter 4. Interdividual Psychology

1. Arturo Pérez-Reverte, *Cadix, ou la diagonale du fou* (Paris: Le Seuil, 2011).

2. A word coined by Arthur Koestler from the Greek *holos* (totality) and the suffix *on* (a part).

3. See René Girard, *The One by Whom Scandal Comes*, trans. Malcolm B. DeBevoise (East Lansing: Michigan State University Press, 2014). See in particular chapter 1, "Violence and Reciprocity."

4. See Eugene Webb, *The Self Between: From Freud to the New Social Psychology of France* (Seattle: University of Washington Press, 1993).

Chapter 5. Psychological Time and the Nodal Points N and N'

1. Cesáreo Bandera, *Mimesis conflictiva: Ficción literaria y violencia en Cervantes y Calderón* (Madrid: Gredos, 1975).

Chapter 6. The Three Brains

1. Jeremy Rifkin, interview, *Le Nouvel Observateur*, August 12–18, 2011.

2. Rifkin, interview.

3. Daniel Goleman, *Emotional Intelligence: Why It Can Matter More Than IQ* (New York: Bantam Books, 2005), 10.

4. Goleman, *Emotional Intelligence*, 9.

5. Goleman, *Emotional Intelligence*, 9.

6. Antonio Damasio, *Descartes' Error: Emotion, Reason, and the Human Brain* (New York: Penguin, 2005), 22.

7. Damasio, *Descartes' Error*, 7.

8. Damasio, *Descartes' Error*, 13.

9. Goleman, *Emotional Intelligence*, 234.

10. Damasio, *Descartes' Error*, 100.

11. Damasio, *Descartes' Error*, 235.

12. Damasio, *Descartes' Error*, 240.

13. See Jean-Michel Oughourlian, *Le désir, énergie et finalité* (Paris, L'Harmattan, 1999).

14. On this point I would be happy to have the opinion of my friend Boris Cyrulnik, one of the creators of this concept, which has given hope to patients, to victims, and also to psychotherapists.

15. Jacques Lacan, *De la psychose paranoïaque dans ses rapports avec la personnalité suivi de Premiers écrits sur la paranoïa* (where Lacan mentions these subjects) (Paris: Le Seuil, 1975), 14.

Chapter 7. The Three Possibilities of the Interdividual Rapport

1. Here I would like to clarify an important point: the model become obstacle should not be confused with taboo, rules, limits, prohibitions. Limits help desire to structure itself; they stimulate desire, excites curiosity, and if desire persists despite the difficulty, it is transformed into willpower.

2. The French edition of *Contributions to the Psychology of Erotic Life*.

3. Freud, *Psychologie de la vie amoureuse*, 8.

Chapter 8. Classical Nosology

1. This diagnosis does not exist in American psychiatry, which would places such mental illnesses under the schizophrenia headings. But in the European system, because these psychoses display but a single "theme," traditionally they are not classified as forms of schizophrenia.

2. Henri Ey, *Manuel de psychiatrie* (Paris: Masson et Cie., 1967), 469.

Chapter 9. Figures of the Other in Normal Experience

1. See Michel Leiris, *La possession et ses aspects théâtraux chez les Éthiopiens de Gondar, précédé de La croyances aux génies zâr en* Éthiopie *du Nord* (Paris: Le Sycomore, 1980).

2. Catalepsy is the second phase of hypnosis, which consists in rigid immobility.

3. Gustave Le Bon, *Psychologie des foules* (Paris: PUF, 1963), 14 (my emphasis).

4. Philippe de Felice, *Poisons sacrés, ivresses divines* (Paris: Albin Michel, 1970).

5. Le Bon, *Psychologie des foules*, 23.

6. See Oughourlian, *The Puppet of Desire: The Psychology of Hysteria, Possession, and Hypnosis*, trans. Eugene Webb (Stanford: Stanford University Press, 1991), 211.

7. We need only think of Girard's analyses concerning the relationship between Hölderlin and Schiller, for example.

8. Pierre Corneille, *Le Cid*, trans. Richard Wilbur, in "*Le Cid*" *and* "The Liar" (Mariner Books, 2009).

9. James Joyce, "The Dead," in *Dubliners* (New York: Bantam Classics, 2005), 187.

10. Joyce, "The Dead," 188.

11. Joyce, "The Dead," 188.

12. Joyce, "The Dead," 188.

13. Joyce, "The Dead," 189.

14. Joyce, "The Dead," 189.

15. Jean-Christophe Rufin, *Le grand coeur* (Paris: Gallimard, 2012).

Chapter 10. Figures of the Other in Neurotic Experience

1. Jacques Corraze, *De l'hystérie aux pathomimies* (Paris: Dunod, 1976).

2. Hysterical conversion is a somatization in the form of a functional physical symptom.

3. The patient can stretch out his (or more often her) legs in the bed, but cannot get out of bed or stand up.

4. Robert Musil, *The Man without Qualities*, trans. Sophie Watkins and Burton Pike (London: Picador, 1995), 678–681.

5. Franz Kafka, *The Trial*, trans. Willa Muir and Edwin Muir, revised by E. M. Butler (New York: Alfred A. Knopf, 1964), 267–268.

6. Kafka, *The Trial*, 269.

7. Kafka, *The Trial*, 269.

Chapter 11. Figures of the Other in Psychotic Experience

1. Laure Murat, *L'homme qui se prenait pour Napoléon: Pour une histoire politique de la folie* (Paris: Gallimard, 2011), 18.

2. This of course resonates with René Girard's description of the scapegoat mechanism: the scapegoat is guilty because he is accused, rather than being accused because he is guilty.

3. Lacan, *De la psychose paranoïaque*, 37.

4. Lacan, *De la psychose paranoïaque*, 31.

5. Lacan, *De la psychose paranoïaque*, 37.

6. Lacan, *De la psychose paranoïaque*, 37.

7. Lacan, *De la psychose paranoïaque*, 37.

8. Lacan, *De la psychose paranoïaque*, 37 (my emphasis).

9. Lacan, *De la psychose paranoïaque*, 39.

10. Lacan, *De la psychose paranoïaque*, 109.

11. Lacan, *De la psychose paranoïaque*, 116.

12. Lacan, *De la psychose paranoïaque*, 117 (my emphasis).

13. Recall the PET scan of the monkey that "lights up" if a man or another monkey grasps a banana, but remains mute if it is a robot or a lever arm that makes the gesture.

14. Lacan, *De la psychose paranoïaque*, 211–212 (Lacan's emphasis!).

15. Lacan, *De la psychose paranoïaque*, 289.

16. Jean-Jacques Rousseau, *A Discourse on Inequality*, trans. Maurice Cranston (New York: Penguin, 1985), 109.

17. Rousseau, *A Discourse on Inequality*, 81.

18. Rousseau, *A Discourse on Inequality*, 81.

19. Rousseau, *A Discourse on Inequality*, 31.

20. See Hans Selye, *The Stress of Life* (New York: McGraw Hill, 1956).

21. Rousseau, *A Discourse on Inequality*, 83.

22. Rousseau, *A Discourse on Inequality*, 84.

23. Rousseau, *A Discourse on Inequality*, 89.

24. Rousseau, *Reveries of a Solitary Walker*, trans. Peter France (New York: Penguin Classics, 1980), 27.

25. Rousseau, *Reveries*, 27–28.

26. Molière, *The Misanthrope*, trans. Richard Wilbur (New York: Harcourt Brace, 1965), 18.

27. Molière, *The Misanthrope*, 18.

28. Molière, *The Misanthrope*, 66.

29. Ey, *Manuel de psychiatrie*, 579.

30. *Nietzsche: A Self-Portrait from His Letters*, ed. Peter Fuss and Henry Shapiro (Cambridge, Harvard University Press, 1971), 9.

31. Nietzsche is referring to *The Birth of Tragedy*, which was published in 1872.

32. Fuss and Shapiro, *Nietzsche*, 29.

33. Fuss and Shapiro, *Nietzsche*, 42.

34. Fuss and Shapiro, *Nietzsche*, 44.

35. Fuss and Shapiro, *Nietzsche*, 45.

36. Fuss and Shapiro, *Nietzsche*, 45.

37. Fuss and Shapiro, *Nietzsche*, 46.

38. Fuss and Shapiro, *Nietzsche*, 46.

39. Fuss and Shapiro, *Nietzsche*, 46.

40. Fuss and Shapiro, *Nietzsche*, 47.

41. Fuss and Shapiro, *Nietzsche*, 65.

42. Letter to Peter Gast, in Fuss and Shapiro, *Nietzsche*, 94.

43. Fuss and Shapiro, *Nietzsche*, 105.

44. Fuss and Shapiro, *Nietzsche*, 105.

45. Fuss and Shapiro, *Nietzsche*, 108 (letter to Georg Brandes).

46. Fuss and Shapiro, *Nietzsche*, 123.

47. Fuss and Shapiro, *Nietzsche*, 126.

48. Fuss and Shapiro, *Nietzsche*, 137.

49. Fuss and Shapiro, *Nietzsche*, 138.

50. Fuss and Shapiro, *Nietzsche*, 140.

51. Henri Grivois, *Grandeur de la folie* (Paris: Robert Laffont, 2012).

Chapter 12. Mood Disorders

1. On the subject of lengthy psychoanalysis without results, see François Roustang's illuminating book, *Elle ne le lâche plus* (Paris, Minuit, 1980).

2. Irvin Yalom, *The Gift of Therapy* (New York: HarperCollins, 2002).

Chapter 13. Diseases of Desire

1. René Girard, *Things Hidden since the Foundation of the World*, trans. Stephen Bann and Michael Metteer (Stanford: Stanford University Press, 1987), 326–327.

2. Girard, *Things Hidden*, 329–330.

3. The third wife of the Roman emperor Claudius had a reputation for nymphomania. Her name has become synonymous with multiple sexual conquests.

4. Molière, *Don Juan*, trans. Richard Wilbur (New York: Harcourt, 2001), 19.

5. Marie-France Hirigoyen, *Abus de faiblesse et autres manipulations* (Paris: Jean-Claude Lattès, 2012), 12. See also (in English) *Stalking the Soul*, trans. Thomas Moore (New York: Helen Marx Books, 2000).

6. Hirigoyen, *Abus de faiblesse*, 54.

7. Hirigoyen, *Abus de faiblesse*, 172.

8. Freud, *Contributions*, 254.

9. Freud, *Contributions*, 57.

10. Freud, *Contributions*, 253.

11. Neuburger, preface to Freud, *Psychologie de la vie amoureuse*, 15.

12. Crébillon fils, *Le sopha* (Paris: Flammarion, 1995), 135.

13. Wilhelm Stekel, *La femme frigide* (Paris: Gallimard, 1948), 209.

14. Stekel, *La femme frigide*, 207.

15. Stekel, *La femme frigide*, 107.

16. Stekel, *La femme frigide*, 210.

17. Foreword (trans. Malcolm DeBevoise) to René Girard, *Anorexia and Mimetic Desire*, trans. Mark R. Anspach (East Lansing: Michigan State University Press, 2013), ix.

18. Or women, in the much less frequent cases of masculine anorexia.

19. Girard, *Anorexia and Mimetic Desire*, 28.

20. Jean-Michel Oughourlian, preface to René Girard, *Anorexie et désir mimétique* (Paris: L'Herne, 2008), 11.

21. Jean-Michel Oughourlian, *La personne du toxicomane: Psychosociologie des toxicomanies actuelles dans la jeunesse occidentale* (Toulouse: Privat, 1973), 254.

22. J.-M. Sutter and H. Luccioni, "Le syndrome de carence d'autorité," *Revue de neuropsychiatrie infantile*, March–April 1959, 1–15.

23. Oughourlian, *La personne du toxicomane*, 260.

24. Oughourlian, *La personne du toxicomane*, 260.

25. Oughourlian, *La personne du toxicomane*, 262.

Chapter 14. The Mimetic Mechanism

1. Lacan, *De la psychose paranoïaque*, 296 (my emphasis).

2. Lacan, *De la psychose paranoïaque*, 296.

Chapter 15. Some Clinical Studies

1. Lacan, *De la psychose paranoïaque*, 392.

2. Lacan, *De la psychose paranoïaque*, 392.

Conclusion

1. Lacan, *De la psychose paranoïaque*, 278 (Valéry's emphasis).

2. A term used by Girard, which he himself takes from Christ in the Gospels, to denote the obstacle, the stumbling block.

3. Voltaire, *Candide*, trans. Peter Constantine (New York: Modern Library, 2005), 114. The baron is Cunégonde's brother and opposed the marriage, asserting that his sister would never marry "anything less than a Baron of the Empire."

4. Lacan, *De la psychose paranoïaque*, 228.

5. Voltaire, *Candide*, 118.

6. Voltaire, *Candide*, 118.

7. *Translator's note.* A critique of the French so-called thirty-five hour workweek, interpreted here as a symptom of a bureaucratized and dissatisfying work culture.

8. Voltaire, *Candide*, 116.

9. Lionel Duroy, *Le chagrin* (Paris: J'ai lu, 2011).

10. Duroy, *Le chagrin*, 306.

11. Duroy, *Le chagrin*, 308.

12. Duroy, *Le chagrin*, 429.

13. Stefan Zweig, *Conscience contre violence* (Paris: Le Livre de poche, 2010), 18–19.

14. Zweig, *Conscience contre violence*, 19.

15. See Eckhart Tolle, *The Power of Now* (Vancouver: Namaste, 1997).

Bibliography

Anspach, Mark Rogin. À *charge de revanche: Figures* élémentaires *de la réciprocité*. Paris: Seuil, 2002.

Bandera, Cesáreo. *Mimesis conflictiva: Ficción literaria y violencia en Cervantes y Calderón*. Madrid: Gredos, 1975.

Bernheim, Henri. *Hypnotisme et suggestion: Hystérie, psychonévroses, neurasthénie, psychothérapie*. 3rd ed. Paris: Doin, 1910.

Carroll, Lewis. *Alice au pays des merveilles*. Trans. Jacques Papy. Paris: Gallimard, 1994.

Corneille, Pierre. "Le Cid" *and* "The Liar." Trans. Richard Wilbur. New York: Mariner Books, 2009.

Corraze, Jacques. *De l'hystérie aux pathomimies*. Paris: Dunod, 1976.

———. *Les communications non-verbales*. Paris: PUF, 1992.

Crébillon fils. *Le sopha*. Paris: Flammarion, 1995.

Damasio, Antonio. *Descartes' Error: Emotion, Reason, and the Human Brain*. New York: Penguin, 2005.

———. *Le sentiment même de soi: corps, émotions, conscience*. Paris: Odile Jacob, 1999.

de Felice, Philippe. *Poisons sacrés, ivresses divines*. Paris: Albin Michel, 1970.

de Keukelaere, Simon. "Des découvertes révolutionnaires en sciences cognitives: les paradoxes et dangers de l'imitation." *Automates intelligents* 63 (2005).

Duroy, Lionel. *Le chagrin*. Paris: J'ai lu, 2011.

Eliade, Mircea. *Rites and Symbols of Initiation*. New York: Harper, 1965.

Erickson, Milton, with Ernest L. Rossi. *L'intégrale des articles de Milton H. Erickson. Vol. 2, Altération par l'hypnose des processus sensoriels, perceptifs et psychophysiologiques*. New York: Irvington, 1980.

——. *"L'hypnose profonde et son induction."* In *Experimental Hypnosis: A Symposium of Articles on Research by Many of the World's Leading Authorities*, ed. M. LeCron, 70–114. New York: Macmillan, 1952.

Ey, Henri. *Manuel de psychiatrie.* Paris: Masson et Cie., 1967.

Falret, Jean-Pierre. *Observations sur le projet de loi relatif aux aliénés.* Paris: Adolphe Everat, 1837.

Faure, Henri. *Les appartenances du délirant.* Paris: PUF, 1966.

——. *Hallucinations et réalité perceptive.* Paris: PUF, 1969.

Freud, Sigmund. "Dostojewski und die Vatertötung." In *Gesammelte Werke* 14: 397–418.

——. *Group Psychology and the Analysis of the Ego.* Trans. James Strachey. New York: Norton, 1959.

——. *Psychologie de la vie amoureuse.* Preface by Robert Neuburger. Paris: Payot & Rivages, 2010.

——. *The Psychology of Love.* Trans. Shaun Whiteside. New York: Penguin, 2007.

Fuss, Peter, and Henry Shapiro, eds. *Nietzsche: A Self-Portrait from His Letters.* Cambridge: Harvard University Press, 1971.

Gallese, Vittorio. "Action Recognition in the Premotor Cortex." *Brain* 119 (1996): 593–609.

——."Embodied Simulation: From Neurons to Phenomenal Experience." *Phenomenology and the Cognitive Sciences* 4 (2005): 23–48.

——. "The Intentional Attunement Hypothesis: The Mirror Neuron System and Its Role in Interpersonal Relations." Accessible online at http://www.interdisciplines.org/mirror/papers/1.

——."The Shared Manifold Hypothesis." *Journal of Consciousness Studies* 8, nos. 5–7 (2001): 33–50.

——."The Two Sides of Mimesis." In *Mimesis and Science: Empirical Research on Imitation and the Mimetic Theory of Culture and Religion*, ed. Scott Garrels, 87–108. East Lansing: Michigan State University Press, 2011.

Garrels, Scott. "Imitation, Mirror Neurons, and Mimetic Desire." *Contagion* 12–13 (2006): 47–86.

——, ed. *Mimesis and Sciences: Empirical Research on Imitation and the Mimetic Theory of Culture and Religion.* East Lansing: Michigan State University Press, 2011.

Girard, René. *Anorexia and Mimetic Desire.* Trans. Mark R. Anspach. East Lansing: Michigan State University Press, 2013.

——. *Deceit, Desire, and the Novel.* Trans. Yvonne Freccero. Baltimore: Johns Hopkins University Press, 1965.

——. *The One by Whom Scandal Comes.* Trans. Malcom DeBevoise. East Lansing: Michigan State University Press, 2014.

——. *Things Hidden since the Foundation of the World.* Trans. Stephen Bann and Michael Metteer. Stanford: Stanford University Press, 1987.

Goleman, Daniel. *Emotional Intelligence: Why It Can Matter More Than IQ.* New York: Bantam Books, 2005.

Grivois, Henri. *Grandeur de la folie.* Paris: Robert Laffont, 2012.

Hirigoyen, Marie-France. *Abus de faiblesse et autres manipulations.* Paris: Jean-Claude Lattès, 2012.

——. *Stalking the Soul.* Trans. Thomas Moore. New York: Helen Marx Books, 2000.

Iacoboni, Marco. *Mirroring People: The Science of Empathy and How We Connect with Others*. New York: Farrar, Straus and Giroux, 2008.

Janet, Pierre. *Névroses et idées fixes*. Alcan, 1898.

Joyce, James. "The Dead." In *Dubliners*. New York: Bantam Classics, 2005.

Kafka, Franz. *The Trial*. Trans. Willa Muir and Edwin Muir. Revised by E. M. Butler. New York: Alfred A. Knopf, 1964.Keysers, Christian. *The Empathic Brain: How the Discovery of Mirror Neurons Changes Our Understanding of Human Nature*. N.p.: Social Brain Press, 2011.

Lacan, Jacques. *De la psychose paranoïaque dans ses rapports avec la personnalité suivi de Premiers* écrits *sur la paranoïa*. Paris: Le Seuil, 1975.

Le Bon, Gustave. *Psychologie des foules*. Paris: PUF, 1963.

LeDoux, Joseph. *Le cerveau des* émotions. Paris: Odile Jacob, 2005.

Leiris, Michel. *La possession et ses aspects théâtraux chez les Éthiopiens de Gondar, précédé de La croyances aux génies zâr en* Éthiopie *du Nord*. Paris: Le Sycomore, 1980.

Maloney Clarence, ed. *The Evil Eye*. New York: Columbia University Press, 1976.

Meltzoff, Andrew N., and Rechele Brooks. "Eyes Wide Shut: The Importance of Eyes in Infant Gaze-Following and Understanding Other Minds." In *Gaze-Following: Its Development and Significance*, ed. Ross Flom, Kang Lee, and Darwin Muir, 217–241. Mahwah, N.J.: Erlbaum.

Meltzoff, Andrew N. and M. Keith Moore. "Explaining Facial Imitation: A Theoretical Model." *Early Development and Parenting*, Vol. 6, 179–192.

———. "Imitation, Gaze, and Intentions." In Scott Garrels, *Mimesis and Sciences: Empirical Research on Imitation and the Mimetic Theory of Culture and Religion*. East Lansing: Michigan State University Press, 2011: 55–74.

———. "Imitation of Facial and Manual Gestures by Human Neonates." *Science* 198, no. 4312 (October 7, 1977): 75–78.

Mesmer, Franz Anton. *Le magnétisme animal*. Paris: Payot, 1971.

———. *Mémoire sur la découverte du magnétisme animal*. Paris: Allia, 2006.

Mitchell, Stephen. *Relational Concepts in Psychoanalysis: An Integration*. Cambridge: Harvard University Press, 1988.

Molière. *Don Juan*. Trans. Richard Wilbur. New York: Harcourt, 2001.

———. *The Misanthrope*. Trans. Richard Wilbur. New York: Harcourt Brace, 1965.

Murat, Laure. *L'homme qui se prenait pour Napoléon: Pour une histoire politique de la folie*. Paris: Gallimard, 2011.

Musil, Robert. *The Man without Qualities*. Trans. Sophie Watkins and Burton Pike. London: Picador, 1995.

Oughourlian, Jean-Michel. *Le désir, énergie et finalité*. Paris: L'Harmattan, 1999.

———. "Desire Is Mimetic: A Clinical Approach." *Contagion: Journal of Violence, Mimesis, and Culture* 3, no. 1 (1996): 43–49.

———. *The Genesis of Desire*. Trans. Eugene Webb. East Lansing: Michigan State University Press, 2009.

————. *La personne du toxicomane*. Toulouse: Privat, 1974.

————. *Psychopolitics: Conversations with Trevor Cribben Merrill*. Trans. Trevor Cribben Merrill. East Lansing: Michigan State University Press, 2012.

————. *The Puppet of Desire: The Psychology of Hysteria, Possession, and Hypnosis*. Trans. Eugene Webb. Stanford: Stanford University Press, 1991.

Palaver, Wolfgang. *René Girard's Mimetic Theory*. East Lansing: Michigan State University Press, 2013.

Pérez-Reverte, Arturo. *Cadix, ou la diagonale du fou*. Paris: Le Seuil, 2011.

Rifkin, Jeremy. *The Empathic Civilization: The Race to Global Consciousness in a World in Crisis*. New York: J.P. Tarcher / Penguin, 2009.

————. Interview with *Le Nouvel Observateur*. August 12–18, 2011.

Rizzolatti, Giacomo, and Corrado Sinigaglia. *Les neurones miroirs*. Trans. Marilène Raiola. Paris: Odile Jacob, 2008.

Rousseau, Jean-Jacques. *A Discourse on Inequality*. Trans. Maurice Cranston. New York: Penguin, 1985.

————. *Reveries of a Solitary Walker*. Trans. Peter France. New York: Penguin Classics, 1980.

Roustang, François. *Elle ne le lâche plus*. Paris: Minuit, 1980.

Selye, Hans. *The Stress of Life*. New York: McGraw Hill, 1956.

Rufin, Jean-Christophe. *Le grand coeur*. Paris: Gallimard, 2012.

Spinoza, Benedict de. *The Collected Works of Spinoza*. Vol. 1. Trans. Edwin Curley. Princeton: Princeton University Press, 1985.

Stekel, Wilhelm. *La femme frigide*. Paris: Gallimard, 1948.

Sutter, J.-M. and H. Luccioni. "Le syndrome de carence d'autorité." *Revue de neuropsychiatrie infantile*, March–April 1959, 1–15.

Thomas, Ben. "What's So Special about Mirror Neurons?" Guest blog available on the *Scientific American* website: http://blogs.scientificamerican.com/guest-blog/2012/11/06/whats-so-special-about-mirror-neurons/.

Tolle, Eckhart. *The Power of Now*. Namaste, Canada, 1997.

Trevarthen, Colwyn, Theano Kokkinaki, and Geraldo Fiamenghi Jr. "What Infants' Imitations Communicate: With Mothers, with Fathers, with Peers." In *Imitation in Infancy*, ed. Jacqueline Nadel and George Butterworth, 127–185. New York: Cambridge University Press, 1999.

Van Eersel, Patrice, ed. *Votre cerveau n'a pas fini de vous étonner*. Paris: Albin Michel, 2012.

Voltaire. *Candide*. Trans. Peter Constantine. New York: Modern Library, 2005.

Watzlavick, Paul, Janet Beavin, and Don Jackson. *Pragmatics of Human Communication*. New York: Norton, 1967.

Webb, Eugene. *The Self Between: From Freud to the New Social Psychology of France*. Seattle: University of Washington Press, 1993.

Yalom, Irvin. *The Gift of Therapy*. New York: HarperCollins, 2002.

Zweig, Stefan. *Conscience contre violence*. Paris: Le Livre de poche, 2010.

Index

A

Adam and Eve, 45–46, 76–77, 85–86, 145, 188, 193
adorcism, 34, 198
agenesis, 159–160
amnesia, 75–t76
anorexia, xxiii, 155–157, 161
anxiety, 92–93
"appropriative mimesis," 44
Aretaeus of Cappadocia, 94
Aristotle, 24, 28, 49, 117
asylums, 107–108
attachment theory, xx
Augustine, 95
autism, 25

B

Bandera, Cesáreo, 44
Bateson, Gregory, 124
Baudelaire, Charles, 127
Bentham, Jeremy, 52
Bernard de Chartres, 82
bipolarity, 133–134, 136
Bowlby, John, xx
Brahms, Johannes, 81
brain functions: first (cognitive), xvii, 49–51, 149, 153; second (emotional), xviii, 51–53, 86, 90, 93, 131, 148–149; third (mimetic), 54–56, 92–94, 98, 99, 138–140, 143, 145, 153. *See also* mirror neuron system; "three brains" theory
bulimia, 155, 157, 161
Bustany, Pierre, 51

C

Carroll, Lewis, 106
Cervantes, Miguel de, 3, 20, 105, 188
children and imitation, 23–25, 28, 34, 54, 77
Christianity, 86–87, 90, 91–92, 109, 129
Clausewitz, Carl von, 100, 152
Coeur, Jacques, 85
cognitivism, 190
conscience, 86–87, 194, 201n1
coquetry, 8
Colbert, Jean-Baptiste, 85
collective intelligence, 62–63
Corneille, Pierre, 83
Corraze, Jacques, 91
cortex, xvii, xviii, xx, 26, 29, 50
Crébillon, Claude-Prosper Jolyot de (fils), 151–152
crowd psychology, 74–76, 166–167
Cyrulnik, Boris, xxii

D

Damasio, Antonio, xviii, 51, 52–53, 109

Darwinism, 115, 117
de Clérambault, Gaëtan Gatian, 66n2, 120
de Felice, Philippe de, 75
delusion, 65–67, 89, 91, 104–109, 111–114, 131,
     140–141, 169; Nietzsche and, 128–129
Descartes, René, xvii, 49–50, 90
desire, 3–8, 11, 54–55, 81, 139, 148, 158–160, 198. *See
     also* mimetic desire; perversions
difference, 55, 115
"diseases of desire." *See* perversions
Don Juanism, 21, 147–150
Dostoevsky, Fyodor, 5, 6, 8, 21, 86–87
drives, xxi, 46
drug addiction, 157–161
DSM (*Diagnostic and Statistical Manual of
     Mental Disorders*), 130, 137, 190
Durkheim, Émile, 141
Duroy, Lionel, 194
dysmorphophobia, 120

E

echolalia, 78
Eersel, Patrice van, 27
emotional intelligence, xviii, 52–53, 109, 117
empathy, xviii, xix, 27–28, 51–52, 62–63, 167
endogenous and exogenous disorders, 133–138
Erickson, Milton, 79, 167, 189, 195
Euler, Leonhard, 35
Ey, Henri, 68, 121

F

Flaubert, Gustave, 4, 6
forgetting, 42–43, 45, 51, 58, 73, 74, 76–77, 190
Fouquet, Nicolas, 85
Freud, Sigmund, 52, 57, 77, 82, 146, 197, 199; on
     conscious and unconscious memory, 51,
     201n2; on the death instinct, 6–7, 52; on
     memory, xvii, 51; on mimetic desire, 20–22,
     203n23; on the self, 54–55; on the superego,
     86–87
     WORKS: *Beyond the Pleasure Principle*, 6,
     52; *Contributions to the Psychology of Erotic
     Life*, 20, 150–152
frigidity, 151–155

G

Gage, Phineas, 52–53
Gallese, Vittorio, xix, 26–27, 109
Garden of Eden myth. *See* Adam and Eve
Gautier, Jules de, 4

Gernsbacher, Morton Ann, 25
Girard, René, xix–xx, xxii–xxiii, 3–4, 33, 54, 109,
     131; on anorexia, 156; on "appropriative
     mimesis," 44; on Freud and Lévi-Strauss,
     82; on homosexuality and perversions, 145;
     pessimism and, 80; on sex, 152. *See also*
     mimetic desire; scapegoat mechanism
     WORKS: *Anorexia and Mimetic Desire*,
     156; *Deceit, Desire, and the Novel*, xxii, 3–9,
     33; *One by Whom Scandal Comes, The*, 36;
     *Things Hidden since the Foundation of the
     World*, xxiii, 6–7, 33, 188
Goleman, Daniel, 52–53, 192
Grivois, Henri, 131

H

hatred, 4, 18
Hippocrates, 94
Hirigoyen, Marie-France, 150
Hobbes, Thomas, 52, 116
Hollande, François, 160, 167, 168
hypnosis, 34, 37–38, 74, 78, 79, 167, 195–198,
     206n2 (chap. 9)
hysteria, 34, 51, 68, 108, 121, 130, 139, 144; attacks
     of, 93, 94–95; hysterical conversion, 93,
     206n2 (chap. 10); in Musil, 97–98

I

Iacoboni, Marco, 27–28
identification, 44, 55, 73, 77, 82, 184–185. *See also*
     models
imitation. *See* mimetic desire; mimetic rivalry
impotence, 150–152, 155
initiation, 6, 151, 159, 188, 196–198, 202n14
interdividual psychology, 18, 33–38, 97, 145, 171,
     176, 180–181, 199; hypnosis and, 195–196;
     interdividual rapport, 18, 54, 56–63, 94, 97;
     sex and, 152–153

J

Janet, Pierre, 78, 98, 196
Janov, Arthur, 195
jealousy and envy, 11–12, 15, 21, 83, 115, 149, 151,
     170; mimetic nature of, 4, 9
Jeanne des Anges, 95, 130
Joyce, James, 83–84

K

Kafka, Franz, 100–101
Keysers, Christian, 27

Kraepelin, Emil, 112
Kretschmer, Ernst, 112
Krishnamurti, Jiddu, 60, 189

L
Lacan, Jacques, 56, 98, 112, 113–114, 168–169, 180, 188
Laclos, Pierre Choderlos de, 147, 149
laughter, 122, 195
Le Bon, Gustave, 75–76
LeDoux, Joseph, 52, 53
Lefort, Guy, 33, 54
Leonardo da Vinci, 43–44, 49
Lévi-Strauss, Claude, 82
limbic system, xviii, xix–xx, 29, 51, 53, 58
Littell, Jonathan, 87
Locke, John, 52
love, three types of, 12–14

M
Machiavelli, Niccolò, 8
manic depression. See endogenous and exogenous disorders
manic excitation, 113, 134, 179
masochism. See sadomasochism
Materazzi, Marco, 27–28
mediation, 3–7, 45, 86; double, 7–8
medication, 79, 138, 139, 166, 189–190
meditation, 198
Meltzoff, Andrew N., 23–25, 33, 34, 109
memory, xvii, 40–43, 50–51, 53, 110
"mental automatism," 66, 120–121, 130–131
mercy, 19
Mesmer, Franz Anton, xx, 34–35, 36, 122
Messalinism, 21, 146–150, 154, 209n3
"metaphysical desire," 6
mimetic contagion, 8, 116–117
mimetic desire, xviii–xx, xxii, 38, 39, 46–47, 58–59, 166, 188; experimental verification of, 23–29; four forms of mimesis and, 43–45; in Genesis, 193; Girard on, 3–11; hypnosis and, 34; interdividual rapport and, 57, 61; jealousy and, 15; Rousseau and, 115–117; the self and, 54, 55; in Spinoza and Freud, 17–22. See also "universal mimesis"
mimetic interdividuality. See interdividual psychology
mimetic nosology, 71
mimetic rapport, xxiii, 76
mimetic psychopathology, xix, 81, 107, 129–131,

150, 171, 189–191, 199; mood disorders and, 138–141
mimetic reciprocity, xviii, xxiii, 36–37, 169, 195
mimetic rivalry, xxi, xxii, 10, 83, 119, 167–169, 172, 176; anorexia and, 156; Candide and, 191–193; desire and, 20–21, 46–47, 58–59; mediation and, 3, 5, 7–8; possession and, 14
mind-body dichotomy, 49–50, 90
mirror neuron system, xviii–xx, 26–29, 145, 190, 204n8; interdividual rapport and, 56, 168–169; second brain and, 51, 52; third brain and, 195
mirrors, 120
Mitchell, Stephen A., xx
models, xxiii, 3–8, 29, 38, 43–45, 54, 57–61, 68, 147; in Duroy, 194; fashion, 155–156; Lacan on prototypes, 168–169; others as, 73–74, 76–77, 80, 84, 87, 89–131, 166. See also mediation; otherness
Molière, 118–119, 147–148
mood disorders, 113, 133–141
Murat, Laure, 108
Musil, Robert, 97–98
mythomania, 89, 90–91
myths, xiv, 47, 118

N
Napoleon I, 87, 100, 103–104, 106, 107, 108–109
Neuburger, Robert, 20, 57, 151
neuroses, 67–69, 92–93, 98–99; religious, 90
Newton, Isaac, 82
Nietzsche, Friedrich, 4, 124–129, 139
nosology of mental disorders, 65–69, 78, 107, 113–114, 159, 180–181, 206n1; third brain and, 71, 98

O
obsessive-compulsive disorders, 99–101
obstacles. See under otherness
Oedipus, 44, 86, 118
otherization of body parts, 93–95, 20–21, 130
otherness, 55, 57, 73–75, 130–131, 187–188; desire and, 39, 117, 199; extrapsychic, 95; Gallese on, 27; historical discussions of, 94–95; intraphysical, 93–95; intrapsychic, 95; as obstacle, 57, 59, 61, 77, 84–87, 98–101, 110, 119–120; recognition and, 42, 45–47; rivalry and, 80–81, 83–86, 92–93, 110–111, 113. See also models

P

paranoia, 66, 111–114, 119, 124, 130, 169, 180;
　　Rousseau and, 114–115, 117–118
paraphrenia, 103–107, 109
pathomimia, 89–90, 91
perversions, xxiii, 7, 143–150; Girard on, 145
physical vs. psychological time, 39–43, 103, 110,
　　198
Piaget, Jean, 24
Plato, 43, 51, 81, 94
"plural somnambulism," 75
possession, 14–15, 34, 51, 109, 130
premature ejaculation, 152
prohibitions and taboos, 5, 86, 118, 157–160, 205n1
Proust, Marcel, 5–6, 40, 43, 53, 202n14
psychasthénie, 98
psychoses, 65–67, 111–113, 124, 180, 183

R

Rabelais, François, 166
rapport. *See* interdividual psychology
recognition and misrecognition, 39, 42–43, 45,
　　92, 114, 190, 198; in Duroy, 194
relational psychoanalysis, xx–xxii, 61
renunciation, 85, 98
resentment, 4, 9, 53, 81, 124, 179
Reza, Yasmina, 99
Rifkin, Jeremy, 51–52
rites of passage. *See* initiation
rivalry. *See* mimetic rivalry; otherness: rivalry and
Rizzolatti, Giacomo, 25–26
Rocky II, 80
Rossi, Ernest, 79
Rousseau, Jean-Jacques, 114–118, 192, 193
Rufin, Jean-Christophe, 85

S

Sade, Marquis de, 108
sadomasochism, xxiii, 144–146
scapegoat mechanism, 61, 118, 207n2
Scheler, Max, 4
schizophrenia, 113, 120–126, 129, 130, 206n1
Schopenhauer, Arthur, 125, 127
Schultz, J. H., 189, 195
self, the, 54–55, 63, 81, 110, 196; collective, 74–75;
　　"conventional," 112–113; desire and, 160;
　　obstacles and, 119–120; overestimation of,
　　114–115, 140
Selye, Hans, 116
Shakespeare, William, 20, 99, 110–111, 151

skandalon, 191, 210n2
Smith, Adam, 52
Socrates, 189
Solomon's judgment, 168
Spinoza, Baruch, xxii, 17–19, 29, 50
Stekel, Wilhelm, 152–153, 154–155
Stendhal, 4, 5, 6, 8
stigmata, 90, 91
Sutter, Jean-Marie, 158

T

Talleyrand-Périgord, Charles Maurice de, 167
Tarde, Gabriel, xx
"three brains" theory, xix–xx, 19, 46–47, 59–63,
　　78–82, 166–171, 174–183, 195–197; desire
　　and, 160; different teachings and, 189; fear
　　and the first brain, 92; Goleman on, 53;
　　mental illness and, 113; mood disorders
　　and, 133–135, 138–140; neuroses and, 68;
　　paraphrenia and, 107; psychoactive drugs
　　and, 175, 189–190; psychopathology and,
　　108–110; psychoses and, 65; rivals and,
　　110–113

U

unconscious, 51, 75, 79
"universal mimesis," 34–36, 41

V

Valéry, Paul, 188
Vega, Lope de, 147
violence, 5, 8, 12, 28, 44; sex and, 150–152
Voltaire, 191–193

W

Wagner, Richard, 125–129, 139
Waterloo syndrome, 100
Webb, Eugene, 38
Willis, Thomas, 50
wisdom, 188–189, 198
work, 134–135, 192–193

Y

Yalom, Irvin D., 139

Z

Zidane, Zinedine, 27–28
Zola, Émile, 22
Zweig, Stefan, 197